From Breakthrough to Blockbuster

From Breakthrough to Blockbuster

The Business of Biotechnology

Donald L. Drakeman, Lisa N. Drakeman, and
Nektarios Oraiopoulos

OXFORD
UNIVERSITY PRESS

OXFORD
UNIVERSITY PRESS

Oxford University Press is a department of the University of Oxford. It furthers
the University's objective of excellence in research, scholarship, and education
by publishing worldwide. Oxford is a registered trade mark of Oxford University
Press in the UK and certain other countries.

Published in the United States of America by Oxford University Press
198 Madison Avenue, New York, NY 10016, United States of America.

Library of Congress Cataloging-in-Publication Data
Names: Drakeman, Donald L., author. | Drakeman, Lisa Natale, author. |
Oraiopoulos, Nektarios, author.
Title: From breakthrough to blockbuster : the business of biotechnology /
Donald L. Drakeman, Lisa N. Drakeman, Nektarios Oraiopoulos,
Description: New York, NY : Oxford University Press, 2022. |
Includes bibliographical references and index.
Identifiers: LCCN 2021038556 (print) | LCCN 2021038557 (ebook) |
ISBN 9780195084009 (hardback) | ISBN 9780197626306 (epub)
Subjects: LCSH: Biotechnology industries.
Classification: LCC HD9999.B442 D73 2022 (print) | LCC HD9999.B442 (ebook) |
DDC 338.4/76606—dc23/eng/20211004
LC record available at https://lccn.loc.gov/2021038556
LC ebook record available at https://lccn.loc.gov/2021038557

DOI: 10.1093/oso/9780195084009.001.0001

Contents

Figures and Tables

Figures

Tables

Preface

Beginning in the 1970s, a series of scientific discoveries promised to transform the business of developing new medicines. As investors sought to capitalize on the excitement over the Nobel Prize–winning discoveries of monoclonal antibodies and recombinant DNA, the biotech industry ultimately grew to thousands of small companies around the world. Each company was trying to emulate what the major pharmaceutical companies had been doing for a century or more, but without the advantages of scale, scope, experience, and massive resources.

Before long, the pharmaceutical companies also adopted these new biotechnologies, while many entrepreneurial biotech companies branched out into medicinal chemistry, long the province of the pharmaceutical industry. Left without a technological edge, how could a large collection of small drug development companies, most with fewer than fifty employees, compete in one of the world's most research-intensive, breathtakingly expensive, and highly regulated industries?

Harvard Business School Professor Gary Pisano asked this question fifteen years ago and concluded that it was essentially a lost cause. In *Science Business*, he wrote, "Unfortunately, the economics have not worked out and biotechnology firms have moved from the frontier to less risky venues."[1] In light of these results, he urged pharmaceutical companies to continue to focus on internal research for the most innovative new medicines.

In this book, we provide the first major reappraisal of the biotech industry since then, and we reach very different conclusions. Our head-to-head comparison of the research productivity of the biotech and pharmaceutical industries shows that biotech has been both more effective and more efficient. An analysis of all products approved by the Food and Drug Administration (FDA) from 1998 to 2016 indicates that biotech companies created nearly 40% more of the most important medicines for unmet medical needs and did so with much lower overall costs. These breakthrough products have often become blockbusters as well. Most of the ten best-selling medicines in 2019 originated in biotech companies.

---

[1] Gary P. Pisano, *Science Business: The Promise, the Reality, and the Future of Biotech* (Brighton, MA: Harvard Business School Press, 2006), 177.

This book[2] is about the diverse collection of biotech companies that has compiled this impressive record. It also focuses broadly on what we call "the biotech ecosystem"—especially the crucial roles played by academic research, venture capital firms, contract research organizations, the capital markets, and the pharmaceutical companies themselves—in providing the supportive environment enabling the biotech industry to create innovative new drugs with unparalleled efficiency. We then conclude by discussing the likely future directions for the biotech industry.

Normally, when an industry has been disrupted by numerous entrepreneurial challengers, predicting the future involves asking when the industry will consolidate and take on mature characteristics. In this case, however, the biotech industry's R&D track record suggests that the drug development business will resist this common pattern. For the reasons we describe throughout the book, innovative drug development will continue to be most efficient and effective in the hands of a large collection of small companies—provided that the ecosystem that supports them continues to thrive. We therefore close by discussing how national policies involving both healthcare economics and distributive justice will set the future course for the survival of that ecosystem as well as the biotech industry itself.

Finally, as we were concluding our writing, the coronavirus pandemic brought the topic of medical research into daily conversations throughout every part of the globe. Academic laboratories, government institutions, biotech companies, and pharmaceutical firms brought an extraordinary level of dedication and cooperation to solving a worldwide crisis. Within about a year after the COVID-19 virus had been identified, the FDA had granted emergency authorizations for a COVID-specific monoclonal antibody treatment and two mRNA vaccines. Those crucial medical developments had proceeded with unusual speed along the pathway from breakthrough to blockbuster described in this book. Technologies that had emerged from government-funded basic academic research were transformed by small biotech companies into completely novel products, which were then developed with funding

[2] Portions of the book first appeared in Donald Drakeman and Nektarios Oraiopoulos, "The Risk of De-Risking Innovation: Optimal R&D Strategies in Ambiguous Environments," *California Management Review* 62, no. 3 (2020): 42–63. Several paragraphs of Chapter 5 previously appeared in Arthur Neuberger, Nektarios Oraiopoulos, and Donald L. Drakeman, "Lemons, or Squeezed for Resources? Information Symmetry and Asymmetric Resources in Biotechnology," *Frontiers in Pharmacology* 8, no. 338 (June 2017): 1–4, https://doi.org/10.3389/fphar.2017.00338. Portions of Chapter 8 were drawn from Donald Drakeman, *Why We Need the Humanities: Life Science, Law and the Common Good* (New York: Palgrave Macmillan, 2016), reproduced with permission of Palgrave Macmillan. Appendix 1 first appeared as an Insight article in the *California Management Review* in May 2020, https://cmr.berkeley.edu/2020/05/covid-manhattan-project/. Appendix 2 was originally published as Arthur Neuberger, Nektarios Oraiopoulos, and Donald L. Drakeman, "Renovation as Innovation: Is Repurposing the Future of Drug Discovery Research?," *Drug Discovery Today*, 24, no. 1 (January 2019), 1–3. All are used with permission.

from investors and through alliances with large pharmaceutical companies. We are hopeful that our efforts here will help shed light on how difficult and impressive that effort was, and how society's ability to respond to similar crises in the future will require all of the various parts of the biotech ecosystem to remain in good health.

We bring a diverse set of backgrounds to this task. Two of us have primarily been biotech entrepreneurs, starting and running biotech companies in the United States and Europe that have completed IPOs, raised billions of dollars, formed alliances with dozens of companies, and been responsible for creating several new FDA-approved treatments for cancer and other diseases. We have also served as board members, strategic advisors, and venture capitalists, which has allowed us to be closely involved with biotech companies in nearly a dozen countries. One of us is a business school professor who has written extensively about the process of innovation and R&D management, has advised entrepreneurial start-ups, and has worked closely on research projects with numerous executives from the biopharmaceutical industry. Together, we have tried to combine the best academic research with thoughtful insights from our own experiences and from the many industry colleagues who have been kind enough to share their views with us.

Over the years, we have had the opportunity to speak to students in science, engineering, business, and even the humanities at Princeton, Oxford, Cambridge, Notre Dame, and other universities. These students have been interested in learning more about entrepreneurship and biotechnology not only from an academic perspective, but also as a potential career direction. In many respects, this book represents the extended version of those talks we have given—our sense of what someone entering the industry might want to know. At the same time, we have tried to analyze the industry, especially as to the origins of important new medicines, in a number of ways that we believe may provide insights for both practitioners and scholars into how the business of drug development has changed over the last half century.

We owe a debt of thanks to more colleagues and coworkers than we could possibly list here, but we do want to mention the catalytic role of Stefan Scholtes, Richard Mason, and the Cambridge Judge Business School's Centre for Health Leadership and Enterprise in bringing the three of us together and in giving us the chance to try out these ideas in Cambridge classrooms and lecture halls. In addition, Nektarios would like to dedicate this book to his parents, Kostas and Sofia.

About the Authors

Donald Drakeman, PhD, was the founding CEO of the U.S. biotech company that pioneered the development of the checkpoint inhibitor cancer treatments recognized in the Nobel Prize for Medicine 2018. These products, YERVOY and OPDIVO, are being used to treat many different forms of cancer. He is a Fellow in Operations and Technology Management at the Cambridge Judge Business School, Distinguished Research Professor in the Center for Citizenship and Constitutional Government at the University of Notre Dame, and a Venture Partner at Advent Life Sciences. His publications have been cited in numerous patents and by the Supreme Court of the United States. He is a Fellow of the Royal Society of Biology. A graduate of Columbia Law School, he received a PhD from Princeton University.

Lisa Drakeman, PhD, was the founding CEO of one of Europe's most successful biotechnology companies. Under her leadership, the company set numerous financing records, including a record-setting IPO, and inaugurated research programs leading to the new FDA-approved medicines DARZALEX (multiple myeloma) and KESIMPTA (multiple sclerosis). She is a board member of the Medical University of South Carolina Foundation for Research Development. She has received numerous industry honors, including the Sol J. Barer Award for Vision, Innovation, and Leadership. She has been a faculty member at Princeton University and regularly lectures on entrepreneurship at universities in the United States and Europe. She received a PhD from Princeton University.

Nektarios Oraiopoulos, PhD, is the Director of the MPhil Programme in Strategy, Marketing & Operations and an Associate Professor of Operations and Technology Management at the Cambridge Judge Business School of the University of Cambridge. His research on innovation and R&D management has appeared in the leading journals of the field. He has won multiple awards and has been invited to make presentations at both academic and industry conferences. In addition to his academic work, he has advised entrepreneurial start-ups and has worked closely on research projects with numerous executives from the biopharmaceutical industry. He holds a Diploma in Electrical and Computer Engineering from the National Technical University of Athens, Greece, and a PhD in Business Administration from the Georgia Institute of Technology.

1

Following the Map of the Genome

In 2000, two teams of researchers raced furiously toward the scientific achievement and personal glory of being the first to unveil the map of the human genome. Each group employed state-of-the-art biological tools and gargantuan computer systems to sift through the massive amounts of genetic information that could be derived from tweezing apart molecules floating in tiny tubes of blood. One team, headed by Francis Collins, was sponsored by the U.S. government, while the other, led by Craig Venter, was funded by a for-profit venture called Celera, one of the hundreds of young companies in America seeking to find new medicines in the modern science of biotechnology. Collins's government labs had $3 billion and a nearly ten-year head start, but Venter's group had a clever new approach, an entrepreneurial spirit, and an IPO-fueled billion-dollar bank account. In the end, it was a dead heat, leading to what Venter optimistically hoped might be an "intellectual peace treaty" after a period of often bitter and contentious competition.[1] At a carefully choreographed press conference hosted by President Bill Clinton, Collins and Venter simultaneously announced the creation of a rough draft of the map of the entire human genome, what some have likened to the discovery of the book of life.

---

[1] Kevin Davies, John Russell, and John Dodge, "John Craig Venter Unvarnished," *Bio-IT World*, November 12, 2002, 44, accessed online, https://web.archive.org/web/20021212085635/http://www. bio-itworld.com/archive/111202/horizons_venter.html. Venter observes that the peace was short-lived. Maynard Olson published a critique of Celera's data in 2002: Maynard V. Olson, "The Human Genome Project: A Player's Perspective," *Journal of Molecular Biology* 319, no. 4 (June 2002): 931–42. Venter responded that "Maynard Olsen has done what many people have told me is probably the most vile article ever published in the scientific literature"; see Davies, Russell, and Dodge, "Venter Unvarnished," 44. Venter may have exercised his well-known talent for hyperbole, but there is no question that Olsen had more than a few bones to pick with Venter in terms of both science and public relations. Ultimately, Olsen couched his verbal assault on Venter's genome mapping in the context of the "essential role" in science of "the vigorous exercise of public criticism"; see Olson, "Human Genome Project," 941. For more along these lines, see Donald Kennedy, "Not Wicked, Perhaps, but Tacky," *Science* 297, no. 5585 (August 2002): 1237. See also John Sulston and Georgina Ferry, *The Common Thread: A Story of Science, Politics, Ethics and the Human Genome* (Washington, DC: Joseph Henry Press, 2002). One year later, an international consortium of academic groups published the final version of the human genome, an announcement that occurred on the fiftieth anniversary of the discovery of the structure of DNA by James Watson and Francis Crick. See Nicholas Wade, "Once Again, Scientists Say Human Genome Is Complete," *New York Times*, April 15, 2003, https://www.nytimes.com/2003/04/15/science/once-again-scientists-say-human-genome-is-complete.html.

Celera, at the turn of the millennium, was in many ways the very model of a modern major genetics company.[2] It had no medicine to sell and marketed no product other than a database of the genes it found, a tool that could potentially be useful for other companies seeking to begin the long and arduous process of drug development. For access to hot-off-the-genomic-map information, pharmaceutical companies were willing to pay dearly, and Celera recorded sales of $120.9 million for the 2002 fiscal year.[3] Celera shareholders watched the genomic excitement drive their stock price up hundreds of percent over just a few months.

Celera was hardly alone in the miraculous Wall Street run-up. The stocks of several hundred biotech companies seemed to defy the laws of gravity as a rising technology tide lifted them to unprecedented levels, thus creating opportunities for the young companies to raise enormous amounts of capital to invest in developing the medicines of the future. Those medicines for cancer, AIDS, arthritis, heart disease, and other afflictions are already playing a major role in twenty-first-century healthcare, and others are still moving through the development process two decades later. But Wall Street's horizons were not nearly as distant as the length of the testing and approval process required by the FDA and similar regulatory bodies around the world. So not long after it touched $250, Celera's stock fell back to $7—once again, along with the rest of the industry. In the meantime, however, the vital cash needed to fuel the product development efforts invigorated Celera and its thousands of biotech compatriots, all believing that a combination of great science and large amounts of money would put the industry into position to inaugurate what some have called "the century of biology."[4]

Beginning in the mid-1970s, the biopharmaceutical industry has been transformed both by biomedical discoveries and by the entrepreneurial revolution they inspired. Breakthrough scientific discoveries, such as those associated with recombinant DNA and monoclonal antibodies,[5] have made

---

[2] A nod to Gilbert and Sullivan's Major-General Stanley seems appropriate for a company that has used a combination of biology and mathematics to sequence the genomes of humans, fruit flies, mice, and other species. In *The Pirates of Penzance*, Major-General Stanley revels in a comparably broad level of interdisciplinary competence: "I am the very model of a modern Major-General, / I've information vegetable, animal, and mineral, / I know the kings of England, and I quote the fights historical, / From Marathon to Waterloo, in order categorical; / I'm very well acquainted, too, with matters mathematical, / I understand equations, both the simple and quadratical, / About binomial theorem I'm teeming with a lot o' news, / With many cheerful facts about the square of the hypotenuse."

[3] See "Celera Sees Slightly Stronger Revenue, Narrowed Loss in Q4," *GenomeWeb*, July 25, 2002, https://www.genomeweb.com/archive/celera-sees-slightly-stronger-revenue-narrowed-loss-q4 .

[4] Craig Venter and Daniel Cohen, "The Century of Biology," *New Perspectives Quarterly* 14, no. 5 (1997): 26–31.

[5] For an introduction to these technologies, see WhatIsBiotechnology.org. See also Lara Marks, ed., *Engineering Health: How Biotechnology Changed Medicine* (London: Royal Society of Chemistry, 2017), and the overview in Chapter 2.

paradigm-shifting changes in the process of developing new medicines. Medicines based on the new science of biotechnology have offered treatments for some of the most intractable diseases, including many forms of cancer, arthritis, and heart disease. But the biotechnology industry is considerably more diverse than just those companies focusing on biological technologies, and some of the industry's blockbuster products are based on the same kinds of medicinal chemistry that the pharmaceutical companies have been using for many years.

How then can we define what a biotechnology company is? That is a remarkably hard question to answer. It is probably too simplistic just to repeat the industry joke about a biotech company being the result of spontaneous combustion when a stockbroker and a biologist come into contact with each other. Although that appears to be how some biotech companies were founded, the landscape is quite diverse, and highly sophisticated venture capitalists are constantly working to collaborate closely with eminent academic scientists (including numerous Nobel laureates) to create new corporate entities based upon each scientific breakthrough of the modern era. But, however the companies may come into being, in virtually all cases, there are fundamentally two parties to these biomedical marriages—patented (or patentable) research with some potential for practical application in the field of healthcare ("the scientist") and a source of cash that will be used to develop the science into a new medicine ("the stockbroker," or, more precisely, an investor).

The basic theory is straightforward: science plus money equals new medicines; new medicines plus patent protection equals monopoly pricing, which is virtually synonymous with high profits, especially when those medicines can significantly enhance, or even extend, patients' lives. The potential for those high profits in the future drives large valuations for the research-stage companies that are expected to reap those profits, thus encouraging investors to risk their money on the new idea in the first place. Of course, both the money side of the business and the scientific elements are considerably more complicated than this summary suggests, but as we survey the many thousands of different companies making up the biotechnology industry, it is important to bear in mind that no matter how sexy the science or fancy the financing, the industry basically boils down to the combination of money and molecules.

That simple equation does not fully capture the impressive array of complexities and potential pitfalls on the route from scientific to commercial success, however. The industry's dedicated efforts have already led to a considerable number of genuinely important new medicines, but the process is rarely simple or straightforward, and biomedical success seems inevitably to be accompanied by numerous failures along the way. Just 10% of new drug candidates entering human clinical testing ultimately become approved medicines,[6] and, in 2018, nearly 90% of publicly traded biotech companies were unprofitable,[7] despite many billions of dollars of investment since the industry began in the 1970s.[8] Thousands of private biotech companies around the world also remain loss-making, development-stage enterprises. Each one is struggling to find the right combination of scientific and financial resources to take a new medicine through the complex and often disappointing process of acquiring sufficient evidence of safety and therapeutic efficacy to obtain the regulatory approval required for the new product's commercial launch. A number of these companies disappear every year—some are acquired, while others run out of money or suffer product failures—and more companies are formed to try other approaches.

If there is a formula for guaranteed success in this arena, it is not clear that anyone has discovered it yet. But, as we will demonstrate throughout the book, the biotech industry has nevertheless shown an impressive ability to generate more novel and important medicines than the pharmaceutical industry, which has had nearly overwhelming advantages of scale, scope, experience, and resources.[9] This achievement was by no means expected by those following what has tended to be the conventional academic wisdom about the industry. In *Science Business*, a highly influential 2006 book on the biotechnology industry, Harvard Business School Professor Gary Pisano observed that the industry was not fulfilling its promises of breakthrough innovations: "It was supposed to be the entrepreneurial biotechnology firms unshackled from tradition and bureaucracy, that would go where big pharmaceutical companies dared not."[10] Yet, the industry's track record over its first thirty years showed that those glowing predictions had not come to pass: "Unfortunately," he concluded, "the economics have not worked out and biotechnology firms have moved from the frontier to less risky venues."[11] In fact, the central thesis of the book was that the biotech industry's "vast scientific success has yet to translate into financial success or improved R&D productivity."[12] In the end, *Science*

---

[6] See Michael Hay et al., "Clinical Development Success Rates for Investigational Drugs," *Nature Biotechnology* 32, no. 1 (January 2014): 40.

[7] This is based on data obtained from Chris Morrison and Riku Lähteenmäki, "Public Biotech 2018—The Numbers," *Nature Biotechnology* 37, no. 7 (July 2019): 714–21.

[8] See Ernst & Young, *Beyond Borders: Matters of Evidence—Biotechnology Industry Report 2013* (London: Ernst & Young, 2013), 25–26.

[9] See, especially, the discussion in Chapter 7.

[10] Gary P. Pisano, *Science Business: The Promise, the Reality, and the Future of Biotech* (Brighton, MA: Harvard Business School Press, 2006), 184.

[11] Pisano, *Science Business*, 177.

[12] Pisano, *Science Business*, 6.

Business encouraged large pharmaceutical companies to continue to look to their own internal research efforts for the creation and development of the most novel products.[13]

Now that the industry has had another fifteen years to deal with the impressively long and complicated process of creating new medicines, the time has come to reevaluate its record. The 2006 analysis covered the industry from its inception in the mid-1970s to 2004, and up to that year, the industry as a whole appeared to be loss making (or very close to it). Moreover, its R&D productivity seemed to be no better than that of the pharmaceutical industry.[14] A decade and a half later, the industry's fortunes have changed markedly for the better. In 2004, for example, only one product created by a biotechnology company was among the top ten best-selling drugs. That product, erythropoietin for anemia, had global sales of $4 billion. It had been created by biotech company Amgen and then licensed to pharmaceutical giant Johnson & Johnson.[15] By the end of 2019, in contrast, a majority of the top ten products had originated in the biotech industry, and they garnered total worldwide sales of $43 billion.[16] Additionally, the monoclonal antibody known as Humira generated the highest revenues of all products that year and is projected to become the best-selling drug of all time. The original development of Humira was the result of a research and development collaboration between a small biotech company and a large pharmaceutical company.[17]

The biotech industry, as a whole, also turned the financial corner shortly after *Science Business* was published. It recorded its first profitable year in 2008 (showing profits of $400 million on $66 billion in revenues) and has been profitable each year thereafter as well, according to Ernst & Young's influential annual report on the industry.[18] In 2012, the U.S. biotech industry, in the aggregate, recorded $5.2 billion of net income,[19] while in 2015 the number jumped up to $16.3 billion,[20] with further income growth hampered primarily

---

[13] See also Gary P. Pisano, "Can Science Be a Business? Lessons from Biotech," *Harvard Business Review* 84, no. 10 (October 2006): 114–24.

[14] See Pisano, *Science Business*, 115, 121.

[15] See Matthew Herper, "The World's Best-Selling Drugs," *Forbes*, March 16, 2004, https://www.forbes.com/2004/03/16/cx_mh_0316bestselling.html#7bd44e5e8e38. The license–licensor relationship could be contentious. See also "Amgen and J&J Fight It Out over Procrit," *The Pharma Letter*, January 20, 2000, https://www.thepharmaletter.com/article/amgen-and-j-j-fight-it-out-over-procrit.

[16] See Lisa Urquhart, "Top Companies and Drugs by Sales in 2019," *Nature Reviews Drug Discovery* 19, no. 4 (April 2020): 228.

[17] See Elizabeth Glasure, "BioSpace Feature: A Look at Miracle Drug Humira's Journey to Proven Efficacy," *BioSpace*, December 5, 2018, https://www.biospace.com/article/biospace-feature-a-look-at-miracle-drug-humira-s-journey-to-proven-efficacy-/.

[18] See Ernst & Young, *Beyond Borders: Global Biotechnology Report 2009* (London: Ernst & Young, 2009).

[19] See Ernst & Young, *Industry Report 2013*, 24.

[20] See Ernst & Young, *Beyond Borders: Staying the Course—Biotechnology Report 2017* (London: Ernst & Young, 2017), 5.

by the fact that most successful companies (the ones called the "commercial leaders" by the Ernst & Young reports) have frequently been acquired by large pharmaceutical companies.[21] These acquisitions have transferred many billions of dollars of both sales and profits to the large pharmaceutical companies, while eliminating them from the composite picture of the biotech industry.

The biotech industry's track record in research and development, including the degree to which it has contributed to the creation of the riskiest, most novel products and technologies, has also evolved considerably. In the past two decades, the biotech industry has outperformed the pharmaceutical industry in the discovery of the most medically important novel drugs, and it has done so at a substantially lower overall cost. In fact, one of the central goals of this book is to understand why that is the case. To do so, we look into the nature of the companies making up the biotech industry, as well as the characteristics of the broader biotech ecosystem that has played a key role in enabling this wave of startups to achieve such a formidable track record.

Where This Book Is Headed

There are many fascinating stories and debates within the realm of biotechnology—scientific discoveries, clinical trials, ethical dilemmas, patent litigation, media attention, Wall Street triumphs and disappointments, and many others. Each area deserves careful attention and thoughtful analysis, much of which is well beyond the scope of this work, which focuses on the key components (or what might be called the major moving parts) of the business of what is generally described as "the biotechnology industry"—that is, the thousands of small companies around the world that are trying to develop new medicines. (There is no doubt that the science of biotechnology has many applications outside of medicine, from food and agriculture to bioremediation of environmental disasters to many others. Our focus in this book, however, is specifically on the story of how the biotech industry develops new medicines.)

We will begin by describing this unusual industry's scientific foundations as well as the long journey from the "light bulb moment," when a scientist has an idea for a new approach to treating patients with a particular disease, to the

---

[21] For instance, the 2013 Ernst & Young report notes, "Between 2008 and 2012, 24 US companies have been commercial leaders [with revenues exceeding $500 million] for at least one year, but only eight of the 24 have been on the list all five years" (Ernst & Young, *Industry Report 2013*, 27). In almost every other case, the company dropped out because it had been acquired.

discovery of a drug candidate, and, with hard work, luck, and a great deal of expense, regulatory approval. This discussion will include the major elements involved in starting and operating a biotech company and how biotech companies are established and supported by a remarkably diverse ecosystem of academic research, venture capitalists, and other investors, managers (many of which are first-time CEOs), contract research organizations, and corporate partners.

These descriptive chapters create the foundation for the first major reevaluation of the biotechnology industry in nearly two decades. In that analysis, we show how and why a diverse and largely disconnected ecosystem of thousands of different investors, by placing individual bets on thousands of different biotech companies, has been responsible for the creation of more novel and medically important therapeutic products than the well-coordinated processes governed by the far more experienced pharmaceutical industry. At the center of the story is the degree to which a Darwinian survival-of-the-fittest capital marketplace has been a critical component of the industry's success in risky new technological arenas. That diverse and competitive financial arena has enabled the industry to take many more shots-on-goal than the pharmaceutical industry while incurring much lower costs as a result of the financial investors' willingness to shut down failures efficiently and even ruthlessly.

Finally, we end by reflecting on where the industry is likely to be headed in the future, and we conclude that the interaction of healthcare economics with issues of social and political justice will have an impressively large influence on those future directions. On the one hand, the governments of numerous countries make very large annual commitments to support basic life sciences research. That multibillion-dollar commitment to "blue sky" academic research, combined with investors' willingness to provide funding for early-stage entrepreneurial biotech companies that are years, perhaps decades, away from launching a product, has stimulated the economy through the creation of many new jobs and has also led to numerous improvements in healthcare. On the other hand, those improvements have been extremely expensive to achieve, even for the relatively efficient biotech industry. Access to the new medicines resulting from medical research, therefore, has been increasingly expensive. Someone—governments, insurance companies, or patients—must pay for the new products once they have reached the commercial market. How much society will be willing and able to pay for better health is still an open question. The way that question is ultimately answered, we believe, is likely to have a greater effect on the future of this industry than any scientific discovery currently on the horizon.

If we have been successful in these efforts, this book will be a useful resource for management students and scholars, biomedical scientists, venture capitalists and other investors, entrepreneurs, healthcare policymakers, and all others interested in how new medicines come into being, why the process is so breathtakingly expensive, and how small entrepreneurial ventures can compete in one of the world's most expensive and heavily regulated industries. Beyond that, the story of how a vast collection of inexperienced, underfunded, and unprofitable small companies have created more life-changing new medicines for less money than the largest global pharmaceutical firms is well worth reading.

The Chapters

Chapter 2 is a brief overview of the major scientific advances that led to the launch of the biotech industry in the 1970s. The first entrepreneurial life sciences companies were founded primarily to take advantage of two Nobel Prize–winning discoveries: recombinant DNA and monoclonal antibodies. Those two disruptive technologies have generated many companies as well as numerous new medicines, but they have not exhausted the possibilities of combining entrepreneurship with drug development. As we highlight in the chapter, there have been a number of other promising new technologies, although their track record in generating novel and important medicines so far has been a fraction of those of the two major ones. Perhaps more importantly, we show the degree to which these new technologies are not exclusively the province of biotech companies, nor are they the only sources of the industry's successful products. In fact, large pharmaceutical companies have adopted these biological technologies in their own laboratories, while biotech companies have branched out into using the same approaches to medicinal chemistry that have been the mainstay of the pharmaceutical industry's drug development research for a century or more. In short, what distinguishes a modern biotech company from a large pharmaceutical company is not so much a focus on using any particular kind of technology but the fact that it is smaller, newer, less experienced, and usually unprofitable.

Chapter 3 explores the drug development process itself: the long, expensive, and highly risky journey between discovery and the regulatory approval needed to commercialize a new medicine. We describe the principal steps, provide an overview of the regulatory framework, and outline the various elements that contribute to the astronomical cost of the drug development process.[22] To make sense of the most recent figures that put the cost per

---

[22] Because of our industry-wide focus, we have not attempted to detail the massive array of complex issues faced by biotechnology companies that are involved in researching, developing, and commercializing

approved drug at over $2.5 billion, we break it down to its major components. The discussion shows that the key drivers behind the ever-increasing costs are a combination of the cost of capital and the remarkably high failure rates of the past decades. Still, even in the rarely occurring best-case scenario, in which everything works as expected, hundreds of millions of dollars are likely to be required to take the drug development process to a successful conclusion, including the need to overcome the array of regulatory hurdles that lie along the pathway to each country's commercial marketplace.

Given these figures about the cost of drug development, the role of fundraising cannot be overstated. This is the focus of Chapter 4. Since many of the molecules have their origins in government grant-funded research, the chapter will follow the ownership of that technology as it moves from universities and other not-for-profit research organizations into for-profit companies that have the potential to attract the investment capital required to pay for the development of a new product. The chapter then describes that capital-raising process through various venture capital stages and ultimately, for some companies, the initial public offering, or IPO, stage. As the industry has achieved more product development success in the last two decades, the financial markets have taken note of the increasing amount of value created as the companies move through the various development stages. Just before the turn of the twenty-first century, the average IPO netted a biotech company about $29 million.[23] By 2019, that amount had grown to $143 million per company,[24] and in the COVID-19-inspired healthcare rally of 2020, highly promising companies with a product in late-stage clinical trials, even those that have yet to obtain FDA approval, could raise many hundreds of millions of dollars at multibillion-dollar valuations.[25]

a new medicine. For an introduction to the many elements of that process, see, for example, Craig Shimasaki, ed., *Biotechnology Entrepreneurship: Leading, Managing and Commercializing Innovative Technologies*, 2nd ed. (London: Academic Press, 2020), which contains forty-two chapters, each of which covers topics about which books and many articles have been written. See also Françoise Simon and Glen Giovannetti, *Managing Biotechnology: From Science to Market in the Digital Age* (Hoboken, NJ: John Wiley & Sons, 2017); Damian Hine and John Kapeleris, *Innovation and Entrepreneurship in Biotechnology, an International Perspective: Concepts, Theories and Cases* (Cheltenham, UK: Edward Elgar, 2006); and Henrik Luessen, ed., *Starting a Business in the Life Sciences: From Idea to Market* (Aulendorf, Germany: Editio Cantor Verlag, 2003). For an analysis of the economics of drug development, see Patricia M. Danzon and Sean Nicholson, eds., *The Oxford Handbook of the Economics of the Biopharmaceutical Industry* (New York: Oxford University Press, 2012); Stuart O. Schweitzer, *Pharmaceutical Economics and Policy*, 3rd ed. (New York: Oxford University Press, 2018); and Kevin M. Murphy and Robert H. Topel, eds., *Measuring the Gains from Medical Research: An Economic Approach* (Chicago: University of Chicago Press, 2003).

[23] See Cynthia Robbins-Roth, *From Alchemy to IPO: The Business of Biotechnology* (Cambridge, MA: Perseus Publishing, 2000), 144.
[24] See Chris Morrison, "2019 Biotech IPOs: Party On," *Nature Reviews Drug Discovery* 19, no. 1 (January 2020): 6.
[25] See, for example, the Belgian biotech company Argenx. With its most advanced product in late-stage clinical trials, it was able to raise over $750 million in a share offering in 2020 and had a market

Chapter 5 covers collaborations, partnerships, and other strategic alliances between biotech and pharmaceutical companies or between two biotech companies. These alliances typically play critical roles in a biotech company's growth and development, and executives spend a great deal of time seeking to arrange them. There are so many alliances—typically over a thousand each year—that, even in the early days, one scholar called the volume of these interfirm transactions "without precedent in business history."[26] Alliances can provide a range of highly valuable resources for a young company. In exchange for giving the pharmaceutical company the right to commercialize a successful product, the biotech company usually receives much needed "upfront" funding at the start of the transaction. These alliances often also offer access to drug development expertise and resources, as well as a link with a larger company with global reach that will add marketing and sales clout and experience if the product is successful. Moreover, not only will the biotech company typically receive a share of the revenues or profits generated by a successful product, but, in the near term, the mere existence of a deal with a well-known pharmaceutical company can be a positive signal to potential investors that a very knowledgeable partner has kicked the scientific tires.

Although investors have recognized the value of that tire kicking, the academic literature has raised a recurring question surrounding these types of transactions. The issue is whether they suffer from a "lemons problem"—that is, the notion that the company developing the product understands its value so much better than any potential partner that only second-rate products will end up being made available. We show that the lemons problem does not appear to exist in this industry. When overall success rates are as impressively low as they are in the drug development business, it appears to be equally difficult for both parties to predict accurately whether the product will ultimately achieve commercial success.

Chapter 6 looks at the backgrounds of the entrepreneurs who become biotech CEOs. With thousands of biotech companies in business at any given time, it would be very difficult for every one of them to recruit a highly

capitalization of over $10 billion. See "Argenx Raises $750 Million in Gross Proceeds in a Global Offering," News, Argenx, last modified May 28, 2020, https://www.argenx.com/news/argenx-raises-750-million-gross-proceeds-global-offering. For a detailed description of the combination of the "art" and "science" of how venture capitalists and institutional investors value biotechnology companies, see Karl D. Keegan, *Biotechnology Valuation: An Introductory Guide* (Hoboken, NJ: John Wiley & Sons, 2008). For the perspective of angel investors, see Scott Dessain and Scott Fishman, *Preserving the Promise: Improving the Culture of Biotech Investments* (London: Academic Press, 2017).

[26] Frank T. Rothaermel, "Complementary Assets, Strategic Alliances, and the Incumbent's Advantage: An Empirical Study of Industry and Firm Effects in the Biopharmaceutical Industry," *Research Policy* 30, no. 8 (October 2001): 1240.

experienced biotech executive to take the helm. As a result, there is an impressive variety of backgrounds among the women and men who have become the CEOs of these companies. Scientists, physicians, and pharmaceutical executives are present in significant numbers, as would be expected, but so are historians, lawyers, regulatory experts, investment bankers, and a range of other professions. Additionally, since it is not uncommon for students pursuing MBA degrees or PhDs in the sciences to ask us whether they should become biotech entrepreneurs, we have outlined a few thoughts they might consider as they continue to seek an answer to that question. Finally, we offer guidance about a number of the key challenges facing biotech CEOs, including raising the money necessary to fund the company's R&D programs.

In Chapter 7 we compare in considerable detail the output of the biotechnology industry, as a whole, to that of the pharmaceutical industry. It turns out that the nascent biotech industry has been much more effective than the large, well-established companies in developing novel medicines that obtain the "priority review" that the FDA provides for new products that address unmet medical needs. After reviewing the possible reasons for this difference, we conclude that the structure of the biotechnology ecosystem—thousands of small companies competing for funding provided by thousands of independent investors—fosters the high volume of risk-taking that enhances the development of genuinely novel products in an environment characterized by many unknown unknowns.[27]

These results point us toward a very different set of strategy recommendations for the major pharmaceutical companies from the one Professor Pisano offered fifteen years earlier. He focused on the need for integration for "the effective development and application of the technology," including access to " 'high fidelity' information" to be able to tell good ideas from "lemons," as well as the opportunity to take advantage of "long-term cumulative learning." As a result, he recommended that large pharmaceutical companies adopt a vertical integration strategy for "the most novel types of innovation," with the "least innovative drugs being 'procured' on the market for know-how."[28] Our results from the study of new drug development since the time of his analysis show that the biotech industry's many shots on goal trump vertical integration in the development of the most novel products by a considerable margin. We therefore encourage large pharmaceutical companies to orient their most novel discovery research toward adopting a more

---

[27] See, generally, Christoph H. Loch, Arnoud De Meyer, and Michael T. Pich, *Managing the Unknown: A New Approach to Managing High Uncertainty and Risks in Projects* (Hoboken, NJ: John Wiley & Sons, 2006).
[28] Pisano, *Science Business*, 159, 177–78.

decentralized internal decision-making approach as well as embracing a greater reliance on external relationships with biotech companies.

Chapter 8 looks to the future of the biotechnology industry and where the map of the genome may lead during the course of the twenty-first century. The governments around the world that invest the over $100 billion a year in fundamental life sciences research usually do so with the dual goal of promoting the health of their citizens, who benefit from the new medicines, as well as of promoting the wealth of their communities, which results from the well-paying jobs created by the companies developing those medicines. So far, the industry has created hundreds of thousands of jobs and has also shown itself to be not only more effective in creating new drugs, but also more efficient than the large pharmaceutical companies in doing so. Nevertheless, the drug development process can be frighteningly expensive, and it will continue to cost a fortune in the future. As a result, the prices of new medicines have risen to levels unimaginable at the outset of the industry, with the expense of new cancer treatments now routinely over $100,000 for a course of therapy and a genetic treatment for a rare childhood disease costing over $2 million.[29] Because of the pressures exerted by the costs of these new medicines on the healthcare system, the industry's future will be substantially determined by whether policymakers, physicians, and patients believe that the costly new medicines emerging from the industry provide enough value to be worth the continued investment in basic life sciences research.

The Appendices

As we were completing this book, the COVID-19 pandemic catalyzed an unprecedented level of research creativity and collaborative interactions among academic, governmental, biotech, and pharmaceutical scientists. A regular question surrounding this flurry of highly focused biomedical research is whether there should be a national, or international, coronavirus "czar," who would take charge of a twenty-first-century medical version of the Manhattan Project—that is, someone empowered to make sure that the best ideas receive the most resources, and so on. Because our research results tend to point in the opposite direction, two of us teamed up with Christof Loch, Dean of the Cambridge Judge Business School, to combine our thoughts with insights

---

[29] See Rob Stein, "At $2.1 Million, New Gene Therapy Is the Most Expensive Drug Ever," *Shots: Health News from NPR*, May 24, 2019, https://www.npr.org/sections/health-shots/2019/05/24/725404168/at-2-125-million-new-gene-therapy-is-the-most-expensive-drug-ever.

from his in-depth study of the original Manhattan Project. The resulting short essay found in Appendix 1 offers a concise summary of many of the key themes of innovation research that we discuss throughout the book.

Appendix 2 relates to a second topic of considerable interest as we completed the book. So many research projects, in academia as well as in large and small companies, have been abandoned over the years that there is potentially a gold mine of "repurposing" opportunities. The basic repurposing thesis is that a product originally in development for one disease—and for which a number of the R&D stages have already been completed—might be able to treat a different disease, much as the hypertension drug candidate now known as Viagra was serendipitously discovered to have a substantially different medical use than the hypertension treatment Pfizer had originally pursued.[30] Perhaps, according to repurposing advocates, medicines can be developed in less time with much higher success rates than has been the case with completely novel products. To explore the possibilities of repurposing, two of us analyzed the existing data with our colleague, Arthur Neuberger. Appendix 2 discusses the outcome of that study, which is the first to track the success ratio of repurposed products. It suggests that there may well be positive outcomes in the future, but the success rates are unlikely to offer a significant improvement over existing drug development efforts.

[30] See "Viagra: How a Little Blue Pill Changed the World," *Drugs.com*, February 24, 2020, https://www.drugs.com/slideshow/viagra-little-blue-pill-1043.

2

The Molecules

Biotechnology will stand as one of the greatest technological advances of the twentieth century, and its potential to leave an enduring imprint on our lives will continue well into the twenty-first century and beyond. Any of the inventions aspiring to hold the top position on the twentieth-century technological totem pole will face stiff competition indeed—from the previously unimaginable, like the Internet, to the easily imagined but practically unthinkable, such as the man on the moon and the mission to Mars, and everything in between: airplanes and automobiles, faxes and phones, radar and radio. All of these were invented or reduced to common practice during the twentieth century. Biotechnology is somewhat more complicated, however. Human healthcare certainly enjoyed life-saving, dramatic improvements during the twentieth century, with vaccines and antibiotics contributing enormous benefits to our longevity as well as our quality of life (especially in industrialized nations). But the fundamental scientific discoveries that characterize modern biotechnology, such as recombinant DNA technology, monoclonal antibodies, and the mapping of the human genome, have yet to be fully exploited, and newer discoveries, such as antisense technology and gene therapy, may still be in their infancy.

To achieve its remarkable potential to revolutionize human healthcare, biotechnology's twentieth-century technological advancements require the fullness of time; many years, massive investments, and dedicated efforts are necessary to match the sophisticated science of modern medicine with the two highly evolved and opposing forces to which they must relate: the remarkably complex human body and the intractable diseases of our era, including cancer, arthritis, heart disease, hepatitis, Alzheimer's, and AIDS. Developing a novel treatment for these kinds of diseases is extremely difficult, and even when such a product shows evidence of medical promise, it cannot be made generally available to the public without government approval. Under U.S. law, and the laws of the many countries that make up the global market for pharmaceutical products, new medicines must be shown to be both safe and therapeutically effective, often in clinical trials costing hundreds of millions of dollars. By contrast, computer programs and mobile apps can be launched into the marketplace and then later debugged, and new products of all sorts go through test marketing before large-scale market launch.

Not so in new biotechnology-based medicines, many of which are based on proteins, molecules, cells, tissues, and organs. For genetically engineered healthcare products, there are no pilot launches of prototypes and no test marketing in selected cities; instead, biotech medicines run the gauntlet of intensive preclinical experimentation and legally mandated, highly structured human clinical testing, a process that traditionally has cost the pharmaceutical industry many millions of dollars and taken a dozen years or more. The result of all of this work is that most new treatments fall by the wayside in the half a generation or more that it takes to develop them, no matter how well they worked in the laboratory. Today, with thousands of genetically engineered medicines in human clinical trials, we may be in the closest thing to biotech's era of beta testing, but in the matter of living things, it may be more accurate to describe it as biotech's gestation period. These efforts have already borne substantial fruit, and quite a bit more is yet to come as biotechnology is increasingly filling the shelves of our modern medicine chest. Scores of medicines based on the science of biotechnology are now available to patients in the United States and many other countries, and they accounted for over $240 billion in sales in 2018.[1] Throughout the chapter, we will describe a number of important biotechnologies, but we should start with the two that have produced the vast majority of biologics to date: recombinant DNA and monoclonal antibodies.

The Core Biologics: rDNA and mAbs

Among the many blockbusters are two oncology products that may well serve as prototypes for biomedicines. One of the first, originally launched by the iconic biotechnology company Amgen in 1991, is called Neupogen, or Filgrastim. The insight leading to this product originated first in the laboratories of academic researchers who identified the molecule that boosts production of infection-fighting white blood cells known as granulocytes (named for their granular appearance under a microscope).[2]

---

[1] See Evaluate, *EvaluatePharma World Preview 2019, Outlook to 2024*, June 2009, 12, accessed online, https://www.evaluate.com/thought-leadership/pharma/evaluatepharma-world-preview-2019-outlook-2024.

[2] For the background of Neupogen, see David C. Dale, "The Discovery, Development and Clinical Applications of Granulocyte Colony-Stimulating Factor," *Transactions of the American Clinical and Climatological Association* 109 (1998): 27–38; and Graham Molineux, MaryAnn Foote, and Tara Arvedson, eds., *Twenty Years of G-CSF: Clinical and Nonclinical Discoveries* (Basel, Switzerland: Springer Basel, 2012).

Cancer patients undergoing high-dose chemotherapy often experience perilously low levels of these white blood cells. Many cancer drugs work by killing cells that are rapidly growing and dividing, a characteristic common to both malignant tumors and healthy bone marrow cells, where granulocytes and other key elements of the immune system are continuously regenerated. The new molecule discovered by the researchers appeared to encourage the growth of new colonies of the critical white blood cells, so they called it granulocyte-colony stimulating factor (G-CSF). The word "factor" is something of an immunologist's term of art for a molecule that seems to perform an important function in the body, often for reasons not yet fully understood[3]—it could just as easily have been called fertilizer for white blood cells.

The process of figuring out what factors play which particular roles in disease was then, and is now, an extraordinarily important part of medical research and is a great deal harder than it sounds. The immune system contains a phenomenally complex array of interactive signaling devices, and the precise role of any particular molecule is usually very challenging to elucidate. Even as scientists begin to home in on how a molecule behaves under one set of conditions, a single change in any one of a multitude of factors can lead to dramatically different results. Bill Bryson, in his *Short History of Nearly Everything*, describes how time and motion make something as simple sounding as a cell into a frighteningly complex three-dimensional space:

> Blown up to a scale at which atoms were about the size of peas, a cell itself would be a sphere roughly half a mile across, and supported by a complex framework of girders called the cytoskeleton. Within it, millions upon millions of objects—some the size of basketballs, others the size of cars—would whiz about like bullets. There wouldn't be a place you could stand without being pummeled and ripped thousands of times every second from every direction. Even for its full-time occupants the inside of a cell is a hazardous place. Each strand of DNA is on average attacked or damaged once every 8.4 seconds—ten thousand times in a day—by chemicals and other agents that whack into or carelessly slice through it, and each of these wounds must be swiftly stitched up if the cell is not to perish.[4]

Bryson's complex portrait of a cell in full animation provides a glimpse into why it is so difficult to map any particular molecular function. It also should

---

[3] A medical reference encyclopedia does not offer a much more technical definition: "Biologically active substances whose activities affect or play a role in the functioning of the immune system." Reference. MD, s.v. "Immunological Factors," accessed August 28, 2020, http://www.reference.md/files/D007/mD007 155.html.

[4] Bill Bryson, *A Short History of Nearly Everything* (New York: Broadway Books, 2004), 377.

remind us that the line drawings of the immune system that periodically appear in the popular press are but a faint and flat impression of a remarkably dynamic multidimensional process. Many years—perhaps many scientists' careers—could be devoted to attempting to understand how a single molecule, or a family of related molecules, may relate to both healthy and diseased states.

For the future, the map of the genome is increasingly providing a starting point for this type of discovery research, with scientists starting with a genetic sequence found on the map and then working "forward," as it were, from the newly discovered gene toward a point of medical intervention. This pathway has several steps: first, the discovery of the gene, then the molecule encoded by the gene, and ultimately the molecule's biological function. Once the function is illuminated, it may be possible to design a new medicine to manipulate that function, thus affecting the course of the disease.[5] In the earlier days of biotechnology, function typically preceded form—that is, researchers found factors that would influence the behavior of important cells, and then they would work "backward" to find and sequence the relevant gene. Once a potentially important factor had been identified, the next step would be to clone the gene that "codes" for the molecule in question—that is, discovering the gene's DNA sequence, a unique combination of the letters A, C, G, and T that stand for the four components of genes. The DNA sequence is essentially the formula that instructs the body how to make the molecule; it is like the computer software that runs the biological machinery of cells, telling them what protein to make by assembling strings of amino acids in the proper order.[6]

A discovery by scientists at Stanford and the University of California that was patented in 1980 made it possible to take a genetic sequence (the gene that had been cloned) and insert it into an organism, such as a bacterium, whose biological machinery is then essentially "tricked" into producing the human molecule. The best analogy might be a computer virus that inserts itself into your computer's hard drive and runs its own program instead of yours. This process is known as *recombinant DNA* because the relevant strands of DNA in the genes are "recombined" in the proper order in the bacterium, which then becomes a living manufacturing plant for the protein.[7]

[5] See, for example, National Academies of Sciences, Engineering, and Medicine, *Deriving Drug Discovery Value from Large-Scale Genetic Bioresources: Proceedings of a Workshop* (Washington, DC: National Academies Press, 2016).

[6] See, generally, Lara Marks, ed., *Engineering Health: How Biotechnology Changed Medicine* (London: Royal Society of Chemistry, 2017); and Leland H. Hartwell, et al., *Genetics: From Genes to Genomes*, 5th ed. (New York: McGraw Hill, 2014).

[7] See Mark Jones, "Berg, Boyer, Cohen: The Invention of Recombinant DNA Technology," *Medium*, November 11, 2015, https://medium.com/lsf-magazine/the-invention-of-recombinant-dna-technology-e040a8a1fa22.

Once a bacterium or other cell had been engineered to mass-produce the recombinant form of G-CSF, the next step was to see whether this newly discovered biological fertilizer would, in fact, stimulate the growth of white blood cells in the human body the way it had in Petri dishes. Indeed, it did, and after a lengthy process of extensive clinical testing, G-CSF was launched by Amgen in 1991. Since then, legions of patients in countries around the world have benefited from G-CSF's ability to invigorate the immune system. G-CSF accomplished this by restoring large numbers of white blood cells depleted by the chemical poisons that still represent one of our basic lines of defense against most forms of cancer. Amgen's development of G-CSF is just one of numerous examples of the successful use of recombinant DNA technology to generate new opportunities for molecular medicines. But this technology is not the only source of biotech's contribution to the treatment of disease.

Cancer treatment took another step forward in 1999 with the launch of Herceptin, a biotechnology-derived product for women whose breast cancer tumors have a specific molecule on their surface known as HER-2. Herceptin was developed by Genentech, which was perhaps best known for having previously launched biotech's first successful initial public offering (or IPO) back in 1980.[8] Once again, as was the case with G-CSF, the initial discovery leading to the Herceptin product was the identification of a molecule and its genetic code—in this case, a gene and a related protein associated with cancer that was named HER-2. First identified in connection with breast cancer, HER-2 has now been linked with numerous other malignancies. Scientists thought that HER-2 might be a molecule through which a "growth factor" could enter a tumor cell; that is, it could be a "receptor" (essentially a door in the cell wall) for substances circulating in the bloodstream that could nourish the tumor and feed its growth. HER-2 seemed to have the makings of an ideal "target" for tumor therapy since a drug designed around HER-2 could have two chances for success. On the one hand, a product targeting HER-2 might be able to interfere with the tumor's food supply, thus providing an opportunity to stop the growth of the malignancy; on the other hand, a HER-2-targeting drug might kill the tumor cell itself while sparing normal cells that did not have HER-2 on their surface.

Researchers at Genentech and UCLA decided to pursue the HER-2 target with a novel molecular tool, a *monoclonal antibody*, which, along with recombinant DNA technology, ranks as one of the twin foundations of modern

[8] See Robert Bazell, *Her-2: The Making of Herceptin, a Revolutionary Treatment for Breast Cancer* (New York: Random House, 1998). See also Sally Smith Hughes, *Genentech: The Beginnings of Biotech* (Chicago: University of Chicago Press, 2011).

biotechnology. This technology allows scientists to create, in the laboratory, proteins that are virtually identical to the normal antibodies that serve as the eyes and ears of our immune systems. Our bodies are populated with billions of these tiny Y-shaped proteins, which circulate through the bloodstream and tissues on the lookout for viruses, bacteria, and other disease-causing agents. Antibodies are typically preprogrammed to bind to a specific kind of disease, and when they have done so, killer cells are alerted to destroy the antibody-coated germs. Antibodies are thus like the lasers that "paint" a military target, which is then destroyed by a computer-driven smart bomb programmed to explode upon contact with the laser-painted target.

The scientists who first discovered the role of antibodies in bringing about the destruction of pathogens by killer cells employed somewhat different, but equally vivid, imagery. They called the killer cells "phagocytes," from the classical Greek words for "eating" and "cell," because they would engulf and enzymatically chew up a virus or bacterium.[9] An especially large and effective phagocytic cell then was labeled a *macro*phage, or "big eater."[10] A virus coated with many antibodies, and thus marked for elimination by macrophages, is said to be "opsonized," a scientific term derived from the Latin word for relish, like mustard or ketchup, something to make the germ more tasty for the big eaters. Early immunologists clearly had both a sense of humor and a good classical education. Whether we imagine antibodies to be smart bomb components or sandwich toppings, it is clear that antibodies play a crucial role in mobilizing our immune defenses.

Thanks to all of these naturally occurring antibodies, our bodies fend off many diseases every day. Unfortunately, our immune system does not always make antibodies when it should, especially when the disease is caused by one of our own cells, like a cancer cell. Antibodies circulating through the bloodstream conduct surveillance for germs and other foreign invaders, but the immune system distinguishes "self" from "non-self," with antibodies designed to ward off whatever is non-self and to ignore the body's own cells. Sometimes however, self-cells become the cause of disease, as in cancer, thus frequently evading the antibodies' disease-stopping activities.

Scientists in Cambridge, England, first devised a method for making "monoclonal antibodies," or antibodies that could be produced in the laboratory. In particular, they discovered a way to use laboratory mice to produce antibodies that would bind to human cells and then figured out how to duplicate those antibodies in test tubes in large enough quantities to enable their use as

[9] See Siamon Gordon, "Phagocytosis: The Legacy of Metchnikoff," *Cell* 166, no. 5 (August 2016): 1065.
[10] Gordon, "Phagocytosis," 1065.

treatments. The mice could be vaccinated with human cancer molecules, and their immune systems would make antibodies to those cancer cells, not necessarily because the cells were malignant but because they were human and thus looked like foreign invaders to the mouse. The Cambridge scientists, Georges Köhler and César Milstein, who won a Nobel Prize for their discovery, figured out that they could extract antibody-producing cells from the spleens of these immunized mice and fuse them in the laboratory with cells capable of growing indefinitely. The result was a cell that would produce exactly the same antibodies over and over again, so that enough could be made to be used for treating patients.[11]

Only one problem remained: if mice would make antibodies to a target because of its *human*ness, the same process could lead human patients to make their own antibodies to attack the rodent-derived monoclonals because of their *mouse*-ness. That is exactly what happened, making it difficult to treat patients repeatedly with mouse-derived monoclonal antibodies. Thus, it became important to devise methods to make these mouse-derived antibodies more human. Initially, mostly human (or "humanized") antibodies were created by essentially cutting and pasting different bits of a human antibody and a mouse antibody. Numerous humanized antibodies have been approved by the FDA for a range of life-threatening conditions.[12]

More recently, scientists turned their attention toward discovering methods for creating fully human monoclonal antibodies. Gregory Winter of the MRC Laboratory of Molecular Biology in Cambridge, who had already developed the technology for humanizing antibodies,[13] then focused on fully human versions. He was awarded the 2018 Nobel Prize in Chemistry for devising a way to "harness . . . evolution to make medicines" from antibodies derived from large synthetic "phage display" libraries.[14] Humira, the world's best-selling medicine, along with a number of other important new medicines, was created by this method.[15] Meanwhile, Nils Lonberg and colleagues at a small biotech company decided to engineer the mice rather than the antibodies. These "transgenic" mice had their mouse genes for making antibodies eliminated (or "knocked out") and replaced by human genes. As a result, when the transgenic mice were vaccinated with a human disease target, their immune

---

[11] For a history of monoclonal antibody technology, see Lara V. Marks, *The Lock and Key of Medicine: Monoclonal Antibodies and the Transformation of Healthcare* (New Haven, CT: Yale University Press, 2015).

[12] See, generally, Marks, *Lock and Key*.

[13] See Marks, *Lock and Key*, chap. 8.

[14] For more details, see "Sir Gregory P. Winter: Nobel Lecture," December 8, 2018, Nobel Media AB, https://www.nobelprize.org/prizes/chemistry/2018/winter/lecture/.

[15] For a discussion of the broader impact of the technology on medicine and the biotechnology industry, see "Nobel Work That Galvanized an Industry," *Nature Biotechnology* 36, no. 11 (2018): 1023.

systems would create human antibodies.[16] Several therapeutic products, including the immunotherapy cancer treatments recognized in the 2018 Nobel Prize for Medicine or Physiology, were made in these transgenic mice.[17]

Ultimately, humanized antibody technology (that is, the cut-and-paste version) became the foundation for Herceptin, the monoclonal antibody designed to bind to HER-2. In collaboration with the UCLA researchers, Genentech discovered that treatment with the Herceptin antibody could stop tumors from growing and perhaps even cause them to shrink.[18] Approved by the FDA in 1998 for the treatment of breast cancer, Herceptin is thus a model of the new types of genomic medicines of the future. These treatments will be designed to target disease much more precisely than the drugs of the past, virtually all of which are developed along conventional organ-focused categories such as breast, colon, or prostate cancer. Instead of identifying tumors by their location, oncologists are increasingly thinking about what causes them to grow or what markers of malignancy might be used to target them for destruction wherever they occur.

In the case of Herceptin, approximately 20% to 30% of breast cancer patients have tumors bearing significant quantities of HER-2, and it is for those patients—not all women with breast cancer—that Herceptin was initially developed. By first testing patients for the presence of HER-2, oncologists work to predict which patients will have a chance of responding to the medication, thus sparing other breast cancer patients the expense and potential side effects that come with any cancer treatment. Moreover, genetic targets like HER-2 may be involved in other cancers, so in the future, it may become less important to say whether the patient is suffering from lung cancer or kidney cancer than whether the tumor is positive or negative for a particular tumor marker. If it is positive, then a treatment focused on eliminating cells with that marker will be indicated, irrespective of where the tumor originated.

---

[16] See Marks, *Lock and Key*, chap. 8. A number of other transgenic mice have been created by biotech companies to make human antibodies. For a summary of the various approaches to creating therapeutic antibodies, see Ruei-Min Lu et al., "Development of Therapeutic Antibodies for the Treatment of Diseases," *Journal of Biomedical Science* 27, no. 1 (2020): 1–30.

[17] The prize was awarded to James Allison and Tasuku Honjo "for their discovery of cancer therapy by inhibition of negative immune regulation." "The Nobel Prize in Physiology or Medicine 2018," August 2018, Nobel Media AB, https://www.nobelprize.org/prizes/medicine/2018/summary/.

[18] The movie *Living Proof* describes the work of one of the UCLA researchers, Dr. Dennis Slamon, in the development and testing of Herceptin.

Mapping of the Human Genome: Myths and Labyrinths

This kind of targeted approach to cancer treatment, which is revamping how we define and treat disease, requires scientists to identify the new targets to serve as the basis of these smart bombs of the future. The previously unknown disease targets will be located somewhere on the map of the human genome. In mapping our genes and elucidating the structures and functions of the proteins to which those genes relate, scientists are working to discover the HER-2s of the future. The mapping of the genome, however, is not important just because we can finally identify important genes; science has been able to do that for quite some time. What is new is the scale on which genes can be identified. Cloning a gene just a few decades ago was something a scientist might spend years trying to accomplish. In fact, the first gene sequenced by genome mapper Craig Venter, working together with his scientist-spouse Claire Fraser, took a full year using mid-1980s technology.[19] At that rate, mapping a genome might have taken tens of thousands of years of work for one person. Yet just twenty-five years later, the mere sequencing of a gene could be accomplished in hours, perhaps even minutes.

There is, however, no straight line from the discovery of a disease-related gene to a new medical treatment, and the road from a genomic map to the pharmacist's shelf is even more labyrinthine. To help us better comprehend the degree of complexity involved, it may be useful to imagine the results of our research into the genome not so much as a map, but as an old-fashioned telephone book. These directories were typically divided into two separate sections—the white pages, where names are filed alphabetically, and the yellow pages, which group names and numbers into occupational categories such as plumbers, doctors, and florists. We now may have the names of every gene in the genomic telephone book—and these names are filed, as it were, alphabetically, as we might find them in the white pages. But for developing new genomic medicines, a gene's name is not nearly enough information; even the gene's address—liver, heart, kidney—falls well short of the mark. We need to know that gene's function—that is, the biological job it performs in the body. And so, finding a cancer-related gene is a little like trying to locate an oncologist in a New York City phone book. The white pages are not very helpful since cancer doctors could have any last name. What we need is the yellow pages, which assemble and list people by profession. The ongoing work

---

[19] The story is recounted in Ingrid Wickelgren, *The Gene Masters: How a New Breed of Scientific Entrepreneurs Raced for the Biggest Prize in Biology* (New York: Times Books, 2002), 23.

of scientists to identify and catalog potential disease targets is not unlike the immense task of fashioning a user-friendly job-by-job genomic yellow pages from the currently relatively complete and well-organized—but much less useful—white pages.

Cataloging the genomic yellow pages will require legions of biomedical detectives to peer into the deep recesses of tumor tissue, arthritic joint fluids, and other body parts, in each case searching for clues as to whether the expression of a particular gene is "upregulated" (that is, there is more of it) or "downregulated" (less of it) under certain disease settings than would be expected under normal conditions. Some of this work is very much akin to the kinds of bench-top biology done in hospital laboratories to see if a cancer patient's biopsied specimen is malignant or benign. Additionally, scientists are making genetically engineered laboratory animals by "knocking out" specific genes—that is, deleting the particular genes in question from the animal's genome[20]—and then seeing whether the animal with the deleted gene suffers from any particular disease. Now that the mouse genome has been mapped, and it has been determined that "99% of mouse genes have a direct human counterpart," studying mice—a time-honored laboratory task, anyway—has taken on new levels of genomic interest.[21] In trumpeting this research opportunity, the venerable science journal *Nature* noted: "Sorry, dogs—man's got a new best friend,"[22] arguing that the mapping of the mouse genome may deserve "equal billing" with the publication of the human genome, "as it provides the key to unlock the secrets of our own DNA."[23]

Even after scientists have fully cataloged and organized the genomic yellow pages, will the elucidation of tens of thousands of previously unknown genes lead to the cure for all diseases? That is certainly unlikely for many different reasons, but there is no doubt that the future of genomic medicine will provide us with new opportunities to fight disease. To put the identification of roughly 25,000–30,000 human genes in perspective, we need to look at the history of drug development. Scientists estimate that, as of about the year 1990, every single medicine available at that time was based on the knowledge of approximately five hundred disease-related targets. Every pill, injection, vaccination, chemotherapeutic, and the like—all of which made up a nearly multi-hundred-billion-dollar international pharmaceutical

---

[20] Here is a rare instance where the technical term—"knock-out"—is simpler than the explanation.
[21] Alison Abbott, "Genomics: Sorry, Dogs—Man's Got a New Best Friend," *Nature* 420, no. 6917 (December 2002): 729. See also Asif T. Chinwalla et al., "Initial Sequencing and Comparative Analysis of the Mouse Genome," *Nature* 420, no. 6915 (December 2002): 520–62.
[22] Abbott, "Sorry, Dogs," 729.
[23] Abbott, "Sorry, Dogs," 729.

marketplace—was based on the knowledge of approximately five hundred genes and their related proteins.

We certainly cannot expect every gene to be associated with disease; after all, many of those genes will be perfectly normal, healthy parts of whatever makes us human (or, as noted earlier, normal parts of what makes mice and humans close relatives). But scientists have proposed that several thousand of these genes may be related to disease, which means that these genes—and the protein targets they "code for"—will at some point become the raw materials for drug development.[24] Identifying more disease-related genes would create the basis for a growing collection of medicines like G-CSF and Herceptin that will be directed specifically at their genomic targets. Researchers can then apply the modern "tools" of biotech product development—antibodies, recombinant DNA, and so on—to these genomically discovered disease "targets" so as to have the basis for a potential medicine.

From Technologies to Products: The Paradigm-Shifting Biologic Medicines

To understand how biotechnology leads to new medicines, it would be helpful briefly to review the pre–biotech industry background of drug development. The process typically started with compounds: chemicals, or combinations of chemicals, that might have some sort of medicinal effect. Some of these chemicals were fashioned in the laboratory while many were found in nature—fungi, molds, snake venom, and other biological and chemical extracts. The search for a cancer drug might involve assembling libraries of many thousands of these chemicals and then seeing which ones of them would kill cancer cells. As it turns out, finding chemicals that kill cancer cells is not as hard as might be expected; the tricky part is identifying a compound that will kill cancer cells without killing everything else.

Traditional pharmaceuticals were, therefore, not so much designed as they were discovered, in the same sense that we speak of discovering oil, gold, or other precious, naturally occurring resources. Sometimes hidden underground or in tropical rainforests, other times composed of common chemicals, these compounds existed in nature. Pharmaceutical researchers literally dug not only through the periodic table of elements but also under rocks and

---

[24] See Juergen Drews, "Genomic Sciences and the Medicine of Tomorrow," *Nature Biotechnology* 14, no. 11 (November 1996): 1516–18. See also Thomas Reiss, "Drug Discovery of the Future: The Implications of the Human Genome Project," *Trends in Biotechnology* 19, no. 12 (December 2001): 496–99.

tree stumps to build the biochemical libraries from which disease-killing agents might emerge.

In 2000, industry analysts at the well-known McKinsey consulting firm teamed up with their counterparts at then-prominent investment banking firm Lehman Brothers to show where modern genomic "third generation" medicines fit within the history of the pharmaceutical industry over the past century while also looking forward well into the twenty-first century.[25] Their story of medical progress was an evolutionary march from plant extracts and serendipity in the first part of the twentieth century to the high-tech, knowledge-based genomic medicines of the twenty-first century. Whether serendipity in drug development is really only in the past remains to be seen, but there is no doubt that, to the extent that new therapeutic products are constructed on the twin building blocks of disease-associated *targets* and drug development *tools* such as antibodies and recombinant DNA, biotech has made paradigm-shifting improvements in both areas. Through genomics, proteomics, and other discovery research, we are identifying numerous new receptors, factors, and other molecules in various disease pathways that can be targeted for therapeutic development. With those novel targets in hand, biotech companies can apply the modern drug development tools—recombinant DNA, monoclonal antibodies, and the like—to generate potential new treatments.

In our description, which collapses all of drug development into the simple (but admittedly oversimplified) matching of targets and tools, it may seem that the sophisticated science of biotechnology can be summed up by asking one simple question: Do we want either more or less of a targeted molecule? Recombinant DNA can be harnessed to add more of the molecules; monoclonal antibodies can cause them to be eliminated, or at least inactivated. In diseases where the body does not have enough of a particular substance—insulin, human growth hormone, G-CSF, and various other molecules known as interferons and interleukins—recombinant DNA technology can replace or supplement what is missing. Insulin, for example, is no longer obtained from pigs and cows, and human growth hormone no longer routinely comes to us from the pituitary glands of human cadavers. Together with G-CSF and Amgen's erythropoietin, a red blood cell stimulant, these recombinant DNA products accounted for over $6.75 billion of sales in 2012,[26] while the total

[25] McKinsey-Lehman Brothers Report, *The Fruits of Genomics* (New York: Lehman Brothers, 2000).

[26] See Andrew Humphreys, "23rd Annual Report: Top 100 Medicines," *Med Ad News* 32, no. 7 (July 2013): 15, 17, accessed online, https://www.pharmalive.com/wp-content/uploads/2015/02/144812-MedAdNews-July.pdf.

market for recombinant DNA drugs (excluding vaccines) is well over $100 billion.[27]

While recombinant DNA technology can be used to restore various immunological factors, monoclonal antibody technology can be used to eliminate or neutralize substances that the body does not need or want, including disease-causing pathogens and naturally occurring hormones. One monoclonal antibody, Humira, is an immunosuppressive medication used to treat autoimmune diseases such as rheumatoid arthritis, Johnson & Johnson's Remicade blocks a protein that contributes to rheumatoid arthritis, and Roche's Rituxan alerts the body to destroy malignant lymphoma cells. These three products are part of a global monoclonal antibody market that was estimated at $135.38 billion in 2018.[28]

While it would be unfair to reduce the many great advances in biotechnology to a small number of core technologies, it is nevertheless the case that two biotech tools—recombinant DNA technology and monoclonal antibodies—have overwhelmingly been the primary sources of biotechnology products to date.[29] This was true in the industry's early days and remains the case after four decades. In a study of the industry that we will discuss in great detail later in the book, the dominance of recombinant DNA and monoclonal technology is clear: 84 of the 89 FDA-approved products that were based on biological molecules were either recombinant proteins (37 products) or monoclonal antibodies (47 products) (see Figure 2.1).

Despite the technological dominance of monoclonal antibodies and recombinant DNA in the biotech pharmacopoeia, there is certainly much more to biotechnology than using these two particular approaches to turning molecular body parts on and off. New tools continue to be discovered that have the potential to add more therapeutic weapons to the pharmaceutical armamentarium. One such tool is *antisense technology*, which had its first FDA-approved product in 1998. This product, known as Vitavene, is an antisense product (or, in more technical terms, an "antisense oligonucleotide") developed by the biotechnology company ISIS Pharmaceuticals for the treatment

[27] See Biomart, "Market and R&D Analysis of Recombinant Protein Drugs," *Creative Biomart Blog*, June 9, 2017, https://www.creativebiomart.net/blog/market-and-rd-analysis-of-recombinant-protein-drugs/.

[28] See "Global Monoclonal Antibodies (mAbs) Market Report 2020 with Profiles of Johnson & Johnson, Merck, AbbVie, Amgen, GlaxoSmithKline—ResearchAndMarkets.com," *Business Wire*, December 11, 2019, https://www.businesswire.com/news/home/20191211005627/en/Global-Monoclonal-Antibodies-mAbs-Market-Report-2020.

[29] We should point out that the FDA regulates a number of different kinds of biological products that are not included in our discussion of biotechnology, including blood derivatives such as gamma globulin, vaccines, allergen extracts, and the like. For the FDA's definition of "biological products," see U.S. Food & Drug Administration, "What Are 'Biologics' Questions and Answers," February 6, 2018, https://www.fda.gov/about-fda/center-biologics-evaluation-and-research-cber/what-are-biologics-questions-and-answers.

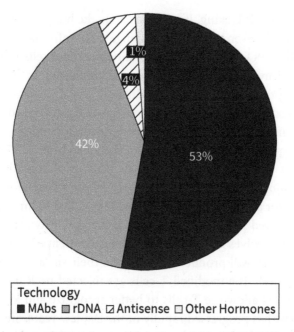

Figure 2.1 Technology of FDA-approved biologic products (1998–2016)
Source: www.fda.gov, EvaluatePharma, annual reports, and patent filings.

of retinitis in cytomegalovirus (CMV) patients whose immune systems have been compromised. The complex mode of antisense activity can perhaps be summarized by saying that it is a very short piece of genetic material that has been designed to block a virus's ability to replicate itself within a cell.[30] As of 2019, antisense technology has produced six FDA-approved medicines.[31]

After antisense, perhaps the most advanced new biological tool is *gene therapy*. Again, the concept is straightforward, even if transforming the basic scientific approach into reality has turned out to be difficult in the years since the first clinical trial in 1990. Once we have cloned the gene for a protein like Amgen's blockbuster erythropoietin (EPO) product, bacteria and other cells can be engineered to produce large quantities of EPO for medical use. If that application of recombinant DNA technology is successful, we should be able to use the same techniques to "program" a patient's own cells to produce the

---

[30] See, generally, Stanley T. Crooke, "Progress in Antisense Technology," *Annual Review of Medicine* 55 (2004): 61–95; and Ruiwen Zhang and Hui Wang, "Antisense Technology," in *Cancer Gene Therapy*, eds. David T. Curiel and Joanna T. Douglas (Totowa, NJ: Humana Press, 2005), 35–49.

[31] Shashwat Sharad, "Antisense Therapy: An Overview," in *Antisense Therapy*, eds., Shashwat Sharad and Suman Kapur (London: IntechOpen, 2019), accessed online, https://www.intechopen.com/books/antisense-therapy/antisense-therapy-an-overview.

EPO that the body is lacking. If bacteria can make human EPO, the patient's own cells should be even better. The challenge that confronts such an approach is in getting the EPO gene programmed into the proper cells and functioning correctly in an environment as dynamic and complicated as the human body. This task has proved to be extremely difficult, and even when it is accomplished, the next critical question will be how to regulate the production of the missing protein so that there is neither too little nor too much.

To date, despite the fact that, as of 2017, there had been 2,597 clinical trials in thirty-eight countries employing some version of gene therapy,[32] gene therapy has enjoyed limited success. In 2017, Novartis's gene-modifying immunotherapy for acute lymphoblastic leukemia, called Kymriah, became the first FDA-approved product based on gene therapy.[33] Kymriah is a CAR-T therapy, which involves genetically modifying a patient's T-cells in the laboratory so that they will produce receptors on their surface (called chimeric antigen receptors). The modified T-cells are injected back into the patient, and the receptors allow the T-cells to better recognize and kill tumor cells. CAR-T therapy has progressed rapidly in the last few years. Shortly after the approval of Kymriah, Gilead's Yescarta (originally developed by the acquired Kite Pharma) received FDA approval for the treatment of adults with large-B-cell lymphoma. In the same year, the FDA also approved Spark Therapeutics's Luxturna for inherited retinal dystrophy that involves inserting a functioning copy of a missing gene directly into a patient's eye (rather than into extracted cells in the lab).

It is also worth noting that the technology used in gene therapy has been far more successful in animals where it has been possible to make genetic changes in the "germ line"—that is, the cells relating to reproduction, thus bringing about an essentially permanent change in the animal's genetic structure for itself and for its progeny. Many scientists around the world have been successful in creating "transgenic" animals, which have some human genes interposed with their normal genes, so that the animal will share certain characteristics with humans, such as bearing a certain kind of human receptor or producing a particular type of human protein. The same approach has been taken to create the "knockout mice" described earlier.

It is not clear whether similar techniques would be as successful in humans as they have been in laboratory animals, but in any event, there have been

[32] See Samantha L. Ginn, Anais K. Amaya, Ian E. Alexander, Michael Edelstein, and Mohammad R. Abedi, "Gene Therapy Clinical Trials Worldwide to 2017: An Update," Journal of Gene Medicine 20, no. 5 (March 2018): 1–16.

[33] Ginn et al., "Gene Therapy," 3. Glybera, another gene therapy–based product that had been approved in Europe in 2012, was withdrawn from the market earlier in 2017.

serious ethical concerns about genetic engineering in human germ lines. Some are worried about the potential for unforeseen consequences for future generations, while others have expressed grave concerns over scientists' attempts to "play God."[34] Aside from these (and other) ethical concerns, there remains a highly challenging technical hurdle to widespread use of gene therapy as a method of treating patients: very few diseases have been linked to a single gene, and even the most promising candidates for gene therapy, such as Huntington's disease or cystic fibrosis, have turned out to be far more genetically complex than previously thought. Without an ability to pinpoint one gene as the "culprit," it is impossibly hard to figure out what genes are to be used in gene therapy. Nevertheless, the CAR-T therapies as well as approved gene therapies, such as Luxturna and Zolgensma (for pediatric patients with spinal muscular atrophy), suggest that the technology continues to have considerable potential.

Tumor vaccines have also made a place for themselves at the biotherapeutic table, although they typically represent something of a misnomer.[35] They do not so much offer the promise of preventing cancer in the fashion that we expect the flu vaccine or the polio vaccine to keep us from contracting those illnesses, as they represent an attempt to use the techniques employed in prophylactic vaccines to create treatments for cancer patients. The goal of conventional vaccines is to mobilize our immune defenses to such a heightened state of immunological preparedness that, upon confronting a disease-causing germ, our bodies will kill or neutralize it before it can establish an infectious foothold. Typically, these normal vaccines contain two major components: (1) a piece of the virus (or a dead virus) that cannot cause illness but will be a good target for the immune system, and (2) an adjuvant, which is basically a substance that will get the immune system excited enough to pay attention to the viral particles (hence the swelling often associated with an inoculation). The immune system picks up the virus and then induces the production of antibodies and disease-fighting cells that are programmed to seek and destroy the virus.

Tumor vaccines seek to employ the same immune mechanism as conventional prophylactic vaccines, but they come into play once a patient has been diagnosed with a malignancy. The immunological hunt is on for tumor-specific molecules that might serve as good candidates for therapeutic vaccines. (In some cases, patient-specific vaccines are being tested in which the patient's

[34] Christopher Gyngell, Thomas Douglas, and Julian Savulescu, "The Ethics of Germline Gene Editing," *Journal of Applied Philosophy* 34, no. 4 (August 2017): 498–513.

[35] See Chunqing Guo et al., "Therapeutic Cancer Vaccines: Past, Present and Future," *Advances in Cancer Research* 199 (2013): 421–75.

own tumor is removed, killed, preserved in the laboratory, and returned in the form of a vaccination.) Once such a molecule has been located, the next goal is to figure out how to stimulate the immune response to a great enough degree to be able to eliminate something as large and established as a tumor, a daunting task when the goal of conventional vaccines is merely to fend off the initial wave of tiny viral invaders when someone with the flu sneezes in your direction. As of 2012, one cancer vaccine, Provenge, for prostrate cancer, has been approved in the United States for the treatment of prostate cancer, several other tumor vaccines have shown promise in preventing recurrence of melanoma and prostate cancer, and hope remains that the insights of genomics will help identify new tumor vaccine candidates.[36]

One often controversial new medical approach is *stem cell therapy*, which seeks to employ versatile cells known as "stem cells" as a source of treatment for neurological and other disorders. Vigorous public debate accompanies every new public release of the potential for stem cell therapy, as President George W. Bush and others have decried the use of fetal or embryonic stem cells, arguing that these potential lives should not be taken in the hopes of finding treatments for others.[37] This ethical debate is likely to continue throughout what will undoubtedly be a very lengthy period during which scientists attempt to decide whether stem cell treatments have the potential for safe and efficacious treatment and whether stem cells harvested from adults or umbilical cord blood might be useful in place of embryonic cells. The most advanced stem cell treatments as of the first decade of the twenty-first century are those such as Prochymal (approved in Canada), which involve the use of adult stem cells to help reconstitute a patient's immune system. In some cases, such as with Prochymal, these are donor-derived adult stem cells; in others, the patients' own cells are used.

In summary, while other biotechnologies are constantly being developed, few, if any, are likely to rival recombinant DNA technology and monoclonal antibodies for their preeminent place in the roster of modern biological medicines, at least in the reasonably foreseeable future. There will almost certainly be additional biotechnologies discovered in the future that may have the potential to contribute to new and improved treatments. But the pathway from inspiration to the first laboratory results to initial product

[36] There are also prophylactic vaccines, such as Gardasil and Cervarit, which are designed to immunize patients against a virus (in this case, human papillomavirus) that may be associated with the development of cancer. When we describe tumor vaccines (or cancer vaccines), we mean antitumor therapies such as Provenge, not these types of vaccines against viral infections that carry a risk of leading to cancer.

[37] The prospect that stem cell therapy may employ cloned fetal cell line rather than donated fetal cells does not necessarily sidestep the ethical and political issues. See Gretchen Vogel, "Scientists Take Step toward Therapeutic Cloning," *Science* 303, no. 5660 (February 2004): 937–38.

candidates—followed almost inevitably by failure and disappointment—then to second-, third-, and fourth-generation approaches, and finally to broad medical and commercial success, may well take decades.

Throughout the decades, the potential of biotechnology research has been advancing. Success in the treatment of life-threatening or seriously debilitating diseases, whether dramatic or modest, is nevertheless impressively time-consuming and expensive. Some group—patients, insurers, or governments—ends up paying for this incremental progress at what often feels like premium prices. One of the questions to which we will return during the course of the book is whether, in light of the time and costs involved, biomedical research and the modern biotechnology industry will continue to enjoy the support of the public, healthcare payers, and investors as the costs of some new medicines for life-threatening conditions often top $100,000 per patient and the cost of medical research around the world exceeds $200 billion per year. But first, we need to complete this tour of the biotechnology industry.

In concluding this scientific introduction, we need to pause to confess that this brief overview of the technological aspects of biotechnology does not do justice to the broad diversity of scientific disciplines that have all given rise to what we call the biotechnology industry. Combinatorial chemistry, high-throughput screening, gene chips, tissue engineering, bioinformatics, proteomics, rational drug design, novel delivery technologies, and the like have given birth to biotechnology companies, and some have directly contributed to creating new medicines as well. Beyond these new biotechnologies, a surprising number of companies widely considered "biotech" companies have been formed around the development of the same kinds of chemical molecules that have been the specialty of large pharmaceutical companies for more than a century. In industry parlance, these pharmaceutically active chemicals are called "small molecules" because recombinant proteins, monoclonal antibodies, and other biologics are much larger and more complex.[38] So, having focused this chapter so far on bio*technologies*, the rest of the chapter will take on a different definitional question: What is a biotech *company*?

---

[38] Chemical drugs and biologics differ in many ways, including their physical size (biologics are orders of magnitude heavier), complexity, molecular structure (chemicals are relatively simple; biologics are typically complex proteins), and manufacture (small-molecule drugs are created through the synthesis of chemicals, whereas large-molecule drugs are protein based and extracted from biological sources and/or produced by biological processes).

What Is a Biotech Company?

Biotech companies are far more diverse than just those enterprises focused on the biological technologies we have just summarized. Take, for example, one of the most prominent leading biotech companies, which is named Celgene. What could be more resonant of the biological medicines of the future than the company's name, which consists of two of biotech's building blocks—"cell" and "gene"?[39] Yet Celgene's first product was not a cell or a gene or even a biologically derived molecule. It was an "old fashioned" small molecule—and not simply any chemical entity; it may be one of the most notorious pharmaceutical products ever developed when it first appeared in the 1950s: thalidomide. Originally used for morning sickness, thalidomide became associated with terrible birth defects. Decades later, Celgene was able to show that thalidomide could be effectively used to initially treat leprosy and then to treat multiple myeloma and other forms of cancer.[40] Building on its expanding knowledge of how thalidomide works, Celgene has subsequently sought to develop a family of drugs that will "modulate" (that is, change or manipulate) the immune system. Celgene, with a massive market valuation, and a very "bio" name, is clearly seen by industry observers and Wall Street as one of biotech's success stories, even if its current association with a cell and a gene was, for quite some time, in name only.[41]

In a similar vein, Vertex Pharmaceuticals is a prominent biotechnology company that has made its name, but in the field of "rational drug design," in which the complex computer modeling of molecules is employed to fashion new treatments from many of the same kinds of chemicals that have historically been used in pharmaceutical products. Vertex's initial success has been chronicled in *The Billion Dollar Molecule: One Company's Quest for the Perfect Drug*.[42] Author Barry Werth's account emphasizes the awesome potential

---

[39] The "Cel" portion of the Celgene name probably derives more from its corporate founder, Celanese Corporation, than from biological cells.

[40] For more on Thalidomide's return as a pharmaceutical product, see Rock Brynner and Trent Stephens, *Dark Remedy: The Impact of Thalidomide and Its Revival as a Vital Medicine* (New York: Basic Books, 2001).

[41] While it has broadened its research focus considerably in the twenty-first century, Celgene indicates on its website that its "major focus" is on "small molecules," rather than large biomolecules, for cancer and immunological diseases. See "A Pipeline of Innovative Therapies to Improve Patient Care," Newsroom, Celgene, last modified January 6, 2014, https://www.celgene.com/pipeline-committed-to-patients/. For more on biotech company names, see Hal Cohen, "A Biotech by Any Other Name," *The Scientist* 17, no. 10 (May 2003), https://www.the-scientist.com/profession/a-biotech-by-any-other-name-51640 , which notes that over fifty companies have names beginning with "gen." A somewhat less obvious choice is monoclonal antibody company, Idec Pharmaceuticals, responsible for Rituxan, one of the industry's most successful cancer products, whose name "came from the Japanese 'Idec Izumi,' which comprises the Chinese characters for harmony and fountain"; see Cohen, "Biotech," 53.

[42] See Barry Werth, *The Billion Dollar Molecule: One Company's Quest for the Perfect Drug* (New York: Simon & Schuster, 1994). Portions of the next two paragraphs first appeared in Donald

of brilliant medical scientists piloting powerful computers through molecular pathways. Vertex's goal was to revolutionize pharmaceutical research by redesigning the drug discovery process. Instead of the time-consuming, labor-intensive process of screening natural substances for medicinal qualities, Vertex would design drugs "rationally." They first identified a compound with great promise, an experimental drug called FK-506, which was used to keep the body from rejecting organ transplants, but which had serious side effects. They then set out to redesign the compound to eliminate its toxic properties by identifying how FK-506 works in the body. Finally, using advanced molecular modeling techniques, they attempted to devise a synthetic product that was even better than the original, which was dug out of the mud on a Japanese mountain.

Vertex ultimately succeeded in designing a replacement for FK-506, a so-called lead compound for animal and human testing. At this climactic moment in the story, however, the readers of *The Billion Dollar Molecule* learn that this potential product did not work as planned; the new compound designed "rationally" by some of the most brilliant scientists of our time was not as good as the original FK-506. Vertex's lead compound, rationally designed to block organ transplant rejection, probably would not do that at all but might just be useful to counteract multi-drug resistance, yet another billion-dollar opportunity. Werth philosophizes that this is just another tenet of rational drug design: that the "ultimate reward for research may turn up elsewhere than intended."[43] It is not clear how rational a design can be if the results are "elsewhere than intended." But while Vertex may have found that it is better to be lucky than to be smart, the hope of rationally designing small-molecule drugs, rather than fashioning large molecules like proteins, continues to serve as the raison d'être of a number of companies widely known as biotech companies.

If the scientific tools and medical and commercial goals of the biotech and the pharmaceutical industries are essentially the same, what is the difference between them? The answer is: not much, perhaps nothing at all. Both industries have employed the same basic tools of drug development: small-molecule chemistry, which has been in use for a century or more, and the newer biological tools.[44] As can be seen in Table 2.1, chemicals, rather than proteins,

---

L. Drakeman's review of Werth's book in Donald L. Drakeman, "Review of *The Billion Dollar Molecule: One Company's Quest for the Perfect Drug,*" *U.S.* 1, August 3, 1994, p. 49.

[43] Werth, *Billion Dollar Molecule*, 416.
[44] See Donald L. Drakeman, "Benchmarking Biotech and Pharmaceutical Product Development," *Nature Biotechnology* 32, no. 7 (July 2014): 621–25; and Marks, *Lock and Key*.

Table 2.1 FDA approvals by technology and company type (1998–2016)

	Small molecules	Biologics
Pharma	269	39
Biotech	171	45
Total	440	89

cells, or other molecular medicines, represent almost 80% (171 out of 216) of the products developed by "biotech" companies. Hence the tongue-in-cheek conclusion reached by one industry observer, usually attributed to investment banker Fred Frank, who said that a biotechnology company is simply a pharmaceutical company "unencumbered by profits."

How then can we define what a biotechnology company is? Many authors at this point might invoke Supreme Court Justice Potter Stewart, who admitted that he could not define pornography but knew it when he saw it; however, inasmuch as the members of the Supreme Court are still struggling to define pornography, no matter how often they see it, the justices are unlikely to help us lay out the metes and bounds of biotechnology. There is no hard-and-fast definition of a biotechnology company based on science, size, or even success in the commercial arena. Instead, we will follow many of the scholars studying the industry, who have generally included within the biotech fold any drug development company founded after 1976, the founding year of Genentech, with pharmaceutical companies defined as any drug development company founded before then.

Choosing Genentech's founding to mark the beginning of the launch of the biotech industry is not just a matter of convenience. When its stock price soared on its first day of post-IPO trading in 1980, the race began for venture capitalists to launch a series of Genentech clones, leading to an industry of thousands of small companies. What essentially defines the industry, then, is that the companies are upstarts, trying to compete in the difficult, often multi-decade-consuming and capital-intensive world of medical research against well-established and well-funded pharmaceutical companies. As we will discuss in subsequent chapters, the biotech industry consists of over five thousand companies around the world, with the vast majority being private (around 86%) and having fewer than fifty employees (around 77%).[45]

---

[45] See Ernst & Young, *Beyond Borders: Returning to Earth—Biotechnology Report 2016* (London: Ernst & Young, 2016), 18; and OECD, *Key Biotechnology Indicators*, December 2011, 1, accessed online, http://www.oecd.org/science/inno/49303992.pdf.

In some industry reports, the term "biotech company" is expanded to include firms applying an array of biotechnology tools to a fairly wide range of commercial applications other than drug development, from genetically modified food to pest control. Nevertheless, we will concentrate on the largest segment of the biotechnology industry—the companies that are devoted to searching for new treatments for human disease. Of all the biotechnology companies in the United States, nearly three-quarters identified themselves at one point as pursuing human healthcare products rather than attempting to explore biotechnology for environmental, agricultural, or other applications.[46] The vast majority of biotech companies reside in this portion of the industry both because the applications—that is, all human diseases—are so rife with potential and because, as bank robber Willie Sutton was rumored to say, that's where the money is. Whatever people might be willing to pay for perpetually red tomatoes, they will pay many times more for the cure for cancer, or even the hope of prolonging a cancer patient's life by a few precious months. That may help explain why it is often the case that nearly half of biotech company products in development are designed to treat cancer, with the remainder focused principally on infectious disease, autoimmune disorders, neurological diseases, and a number of rare diseases.[47]

Much of our focus will also be on the United States, and especially on products approved by the FDA. The United States represents the world's largest pharmaceutical market, and more capital has generally been available there than in other parts of the world. Nevertheless, it is important to point out that various parts of the world have sought to draw entrepreneurs and the jobs they create to their locales. In the United States, biotech companies have often sprouted in hotbeds of medical research like San Francisco and Boston, with a critical mass of companies in a variety of other locales such as San Diego and New Jersey, which calls itself the "nation's medicine chest" because of the large number of pharmaceutical firms in the state.[48] Various state governments have awarded financial incentives to increasing numbers of companies in Maryland, Research Triangle Park in North Carolina, and elsewhere, and nearly all states have had initiatives designed to attract and

[46] See the Department of Commerce's survey, which notes that 747 of the 1,031 companies indicated that human healthcare was their primary focus: U.S. Department of Commerce (USDOC), *A Survey of the Use of Biotechnology in U.S. Industry*, November 2003, 31, accessed online, https://www.bis.doc.gov/index.php/documents/technology-evaluation/25-a-survey-of-the-use-of-biotechnology-in-u-s-industry-2003/file.
[47] See Gary Walsh, "Biopharmaceutical Benchmarks 2014," *Nature Biotechnology* 32, no. 10 (October 2014): 1000, citing data from PhRMA.
[48] John L. Colaizzi, *New Jersey: The Nation's Medicine Chest* (Nutley, NJ: Hoffmann-La Roche, 1985). See also Ernst & Young, *Beyond Borders: Reaching New Heights—Biotechnology Industry Report 2015* (London: Ernst & Young, 2015), 25.

nurture biotech companies.[49] Whether from natural conditions conducive to biotechnology or local programs designed to attract and nurture the industry, it has clearly become a national phenomenon, albeit one focused heavily on the coasts.

These data do not speak for all of North America. Canada, for example, was home to hundreds of biotech companies as of 2019.[50] Similarly, a substantial number of biotech companies have grown up around the United Kingdom's Oxbridge universities, Imperial College, and various other U.K. centers of academic research. Until a high-profile merger took place in 2003, two of the largest English biotech companies had "Oxford" or "Cambridge" in their names, and 291 life science university spin-off companies in the United Kingdom were formed between 2007 and 2011.[51] Biotech entrepreneurship has also extended to essentially all of Europe, from France and Switzerland, to Italy, Germany, and Scandinavia's Medicon Valley, a region comprising parts of both Denmark and Sweden. As of 2018, there were over forty-three hundred biotech companies in Europe.[52]

Other parts of the world have also worked to court biotechnology startups. Australia took the lead in establishing entrepreneurial biotechnology in the Pacific Rim, but Singapore has sought to catch up by initiating an array of public–private initiatives to lure companies to their biocenters. Moreover, biotech companies can now launch IPOs on the stock markets in Japan and China. Worldwide, the revenues of publicly traded biotech companies have grown from just over $11 billion in 1997 to just under $140 billion in 2016.[53]

Conclusion

Originally founded around the scientific promise of novel biotechnologies, especially recombinant DNA (rDNA) and monoclonal antibody (mAb)

---

[49] The Department of Commerce reports that the most rapid growth of biotechnology establishments occurred in North Carolina, which "increased its total number of firms by 52.5% during 1997–2001, up from 23.7% during the previous period." See USDOC, *Survey of the Use of Biotechnology*, 16.

[50] Pharma-Medical Science College of Canada, "Canadian Biotechnology Company List," last modified January 30, 2020, https://www.pharmamedical.ca/canadian-biotechnology-company-list/.

[51] The Oxbridge companies were Oxford GlycoSciences, now part of Celltech, and Cambridge Antibody Technology. See Lord David Sainsbury, "A Cultural Change in UK Universities," *Science* 296, no. 5575 (June 2002): 1929; and Mobius Life Sciences, *Realignment: UK Life Science Start-Up Report 2012* (Nottingham, UK: Mobius Life Sciences, 2012), 6, accessed online, https://biocity.co.uk/wp-content/uploads/pdf/real ignment-uk-life-science-start-up-report-2012.pdf.

[52] See Organisation for Economic Co-operation and Development, "Number of Firms Active in Biotechnology, 2006–2018," last modified October 2020, https://www.oecd.org/sti/KBI1-number-of-firms-active-in-biotech-2020.xlsx.

[53] See Ernst & Young, *Beyond Borders: Staying the Course—Biotechnology Report 2017* (London: Ernst & Young, 2017), 32.

technologies, the biotech industry includes thousands of companies united primarily by the fact that they are upstarts founded after Genentech in 1976 and are pursuing the goal of developing new medicines. The industry had its origins in the United States, but Europe and Asia have raced to catch up, and biotech companies can be found virtually everywhere cutting-edge biomedical research takes place. The disruptive effect on the drug development market from the biotech companies is clear: seven of the top ten marketed products in 2019 had biotech connections.[54]

---

[54] See Kyle Blankenship, "The Top 20 Drugs by Global Sales in 2019," *FiercePharma*, July 27, 2020, https://www.fiercepharma.com/special-report/top-20-drugs-by-global-sales-2019.

3

The Costly Drug Development Process

Fashioning a new pharmaceutical product takes so long and costs so much that it is a wonder it happens at all. A 2013 study from a team of researchers at the Tufts Center for Drug Development, the University of Rochester, and Duke University (the "Tufts Study") pegs the total expense at more than $2.5 billion, an enormous amount of money that needs to be spent over the course of a dozen years or more.[1] Every few years, this study is updated, and the results show ever-increasing costs per product. A study conducted in 2003 by the same authors and using the same methodology estimated the corresponding cost at $800 million in 2000 dollars (which represents $1.044 billion in 2013 dollars[2]), which means that costs increased over 150% in a ten-year period. We will discuss later in this chapter whether the costs shown by those analyses are the right way to think about the R&D process in small biotech companies. For now, suffice it to say that astronomical amounts are required for a biotech company to create, develop, and commercialize a new product before the company becomes financially self-sufficient. Much of this book is about the many ways biotech companies have devised to address their ever-present need for funding. In this chapter, we will try to provide an outline of the overall process and the wide range of R&D approaches biotech companies have taken to advance their technology or products toward successful commercialization.

In considering the full range of R&D activities pursued by the array of four to five thousand companies, it is valuable to bear in mind that most biotech companies are focused on just one or two products or core technologies. Figure 3.1 offers a broad overview of the long journey from the lab to a patient's medicine cabinet. The design of the chart is based primarily on the information provided by the FDA website about the stages of drug development, especially the sections that describe the FDA review process,[3] while the

[1] See Joseph A. DiMasi, Henry G. Grabowski, and Ronald W. Hansen, "Innovation in the Pharmaceutical Industry: New Estimates of R&D Costs," *Journal of Health Economics* 47 (May 2016): 20–33.

[2] See Joseph A. DiMasi, Ronald W. Hansen, and Henry G. Grabowski, "The Price of Innovation: New Estimates of Drug Development Costs," *Journal of Health Economics* 22, no. 2 (March 2003): 151–85.

[3] For more information, see U.S. Food & Drug Administration, "New Drug Development and Review Process," last modified June 1, 2020, https://www.fda.gov/drugs/cder-small-business-industry-assistance-sbia/new-drug-development-and-review-process; and U.S. Food & Drug Administration, "The FDA's Drug Review Process: Ensuring Drugs Are Safe and Effective," last modified November 24, 2017, https://www.fda.gov/drugs/drug-information-consumers/fdas-drug-review-process-ensuring-drugs-are-safe-and-effective.

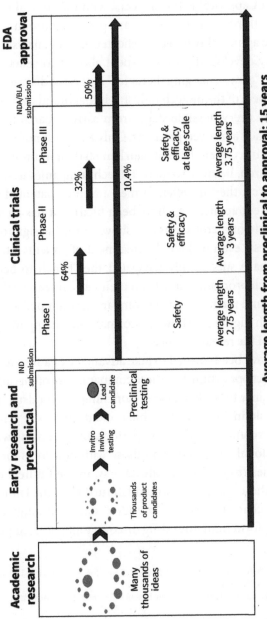

Figure 3.1 The drug development process

Source: www.fda.gov. Michael Hay et al., "Clinical Development Success Rates for Investigational Drugs," *Nature Biotechnology* 32, no. 1 (January 2014): 40–51. Joseph A. DiMasi, Henry G. Grabowski, and Ronald W. Hansen, "Innovation in the Pharmaceutical Industry: New Estimates of R&D Costs," *Journal of Health Economics* 47 (May 2016): 20–33.

estimates about the length and success rates of each clinical phase are derived from the Tufts Study and the most recent major analysis of clinical trial success rates by Hay et al. in 2013,[4] respectively. The average time from preclinical research through FDA approval is fifteen years, with the clinical trials phase occupying about two-thirds of that time. For any particular product, however, the timeline could be appreciably longer or shorter. A 2014 analysis indicated that the most rapidly developed products spent about two years in clinical trials, and those were "game-changing therapies" in serious diseases.[5]

One interesting aspect of the biotech industry is that brand new companies can appear at literally every possible stage, from early research to the commercialization of already approved products. Only a modest number start off with nothing more than a clever idea and then launch their initial efforts at the earliest stages of research. Because so many biotech companies have emerged from academic research, they are often formed around a patented molecule (or one for which patent applications have been filed) that has already been created and subjected to numerous tests that show promising signs of activity. Occasionally, those tests even include the treatment of patients—that is, the academic founders have successfully navigated the "bench-to-bedside" or "translational" research process leading from their basic research labs to their institution's hospital, and the new biotech company has skipped ahead to the "clinical" phase, which Figure 3.1 shows as normally occurring five years (or more) after the start of the research process.

Some companies have been formed to take advantage of a new technology for discovering product opportunities. These companies are often called "platform" companies because they offer a novel launching pad for drug discovery efforts. Examples would include Celera, which sold databases of genomic information; the various companies that have periodically sprung up with a new way to create monoclonal antibodies; and those with new tools for creating large chemical libraries from which products may, in the future, be discovered.[6] In Figure 3.1, these companies would fall further to the left than even the beginning of the chart, that is, the "pre-idea" stage. Although they may have opportunities to form numerous drug discovery alliances, or perhaps to

---

[4] See Michael Hay et al., "Clinical Development Success Rates for Investigational Drugs," *Nature Biotechnology* 32, no. 1 (January 2014): 40–51.

[5] Barbara Bolten, "Fastest Drug Developers and Their Practices," *CenterWatch*, August 1, 2017, https://www.centerwatch.com/articles/13284.

[6] At the other end of the spectrum from platform companies are those biotech companies that have acquired an already-marketed product from a pharmaceutical company in hopes of repositioning it in the marketplace or running further clinical trials to broaden the market. These products have already been shown to be safe and effective. The question is whether they can find a larger and more attractive commercial market. They, too, compete in the marketplace for capital, and they fall all the way on the right side of Figure 3.1.

sell access to their platform technologies, if they decide to use the platform to create their own therapeutic product, they are many years away from having that product reach the FDA approval stage.

Throughout the industry, however, most biotech companies are not focused on a platform; rather, they are product companies—that is, they are dedicated to moving one or more (and rarely more than a few) potential medicines through the lengthy process ending in FDA approval and commercial launch. Even many platform companies grow into being product companies, as they work to employ their own technologies to create a new therapeutic product. These decisions by platform companies to commence product development efforts have regularly provided opportunities for one early-stage biotech company to form an alliance with another. For example, a platform company revolving around identifying new targets for drug development may collaborate with another platform company working on making better tools for fashioning products so as to jointly create a new product based on the combination of their technologies.

From Discovery to Regulatory Approval

Wherever a biotech company initially lands on the R&D spectrum described in Figure 3.1, its goal is to move to the right as rapidly as it can. Prior to initiating human clinical testing, the process may differ considerably based on the nature of the disease and whether the product is a biologic or a chemical entity. Generally speaking, there are initial lab-based in vitro tests to see if the hoped-for activity—killing a cancer cell, blocking a growth factor, neutralizing a virus—actually takes place. If so, these tests are followed by animal (in vivo) tests to see if the product still works under conditions more closely resembling a human patient.

This in vitro and in vivo testing can take years. Unfortunately, the process is usually far more iterative than the linear depiction in Figure 3.1, and it will often involve multiple efforts to reengineer the product. Once this testing yields successful results, the company prepares to file an Investigational New Drug (IND) application with the FDA, which is required to commence human clinical testing. (We should pause to note here that all this work is done with the knowledge that more than 90% of all drugs that failed in human testing had previously passed these preclinical tests with flying colors. If they had not done so, they never would have made it as far as human testing.) For an IND to pass muster, the FDA has a long list of requirements regarding formal preclinical safety testing as well as detailed manufacturing rules to prevent impurities. Even after years devoted to creating and testing a new product, a company will likely spend anywhere from one to more than three years in this pre-IND testing and manufacturing phase. During this time, it will also work with physicians and hospitals to recruit the medical centers that will enroll patients in the clinical trial.

Perhaps the most straightforward portion of the drug development process to describe is the clinical testing stage. Whatever type of molecule has emerged from the research laboratories, whether it is a small-molecule drug or a biologic, it enters a highly structured and heavily regulated clinical development process. The FDA's statutory mandate is fairly straightforward—to determine whether drugs are "safe and effective"[7]—but that simple dictum has evolved into volumes of guidelines for how companies need to demonstrate safety and effectiveness through the clinical trial process. The classic model for clinical development prior to FDA approval consists of three "phases." The first is Phase I, which typically exposes healthy volunteers to the new pharmaceutical agent (usually at various dose levels) to judge both safety and more technical concerns such as the distribution of the drug throughout the body, its duration in the bloodstream, and related questions. Sometimes Phase I trials involve only one dose of the product, even if more doses are expected to be employed in the future.

Products for life-threatening diseases such as cancer sometimes start clinical development in very ill patients who have failed to respond to all existing treatments, rather than in healthy volunteers. These are often called Phase I/II trials because, while they are primarily intended to provide safety data, there is the possibility that they may also show some evidence of therapeutic activity—that is, they may have an effect on the disease by shrinking the tumor or at least slowing the rate at which it is growing. Because the doses are low and the patients are usually at the terminal stage of a cancer that has resisted all known treatments, the likelihood of a positive response is low, although in very few cases there have been impressively good results. One remarkable cancer drug—Novartis's Gleevac—did just that. In a Phase I study, primarily designed to check safety and tolerability, a complete hematological response (meaning that all blood cell counts returned to normal) was observed for 53 out of 54 patients who received daily doses of 300 mg or more, within the first four weeks of therapy.[8] Everyone hopes for a Gleevac-like Phase I trial, but such trials are extremely rare.

---

[7] See U.S. Food & Drug Administration, "New Drug Development and Review Process."
[8] See Brian J. Druker et al., "Efficacy and Safety of a Specific Inhibitor of the BCR-ABL Tyrosine Kinase in Chronic Myeloid Leukemia," *New England Journal of Medicine* 344, no. 14 (April 2001): 1031–37.

In the development of conventional small-molecule drugs, Phase I was often both the alpha and the omega of a product's clinical life span. Many pharmaceuticals have failed to demonstrate a satisfactory safety profile in Phase I trials, and the product is either terminated or sent back for the medicinal chemists to tweak the molecular structure in hopes of fewer side effects. Studies suggest that more than one-third of small-molecule drugs are dropped at the Phase I stage. Not so with biotech products, such as recombinant DNA and antibodies, which have been engineered to be doppelgangers for their natural human equivalents. These kinds of biotech products have generally performed well in safety studies.[9] Because they have a very high likelihood of being well tolerated by patients, they have had a higher chance of successfully emerging from Phase I (or Phase I/II) trials and entering into Phase II testing.

In classical drug development, such as for small-molecule drugs being used to lower cholesterol, Phase II trials have been the first experience of the drug in patients with the relevant condition, and they often include a range of different doses. Safety data continue to be collected, but the primary goal of Phase II is to identify a dose at which enough patients have a positive response (for example, tumor shrinkage, lower cholesterol or blood pressure, or whatever is chosen as the therapeutic "endpoint" of the study) to justify continued development of the product. Whereas Phase I trials can be as small as twenty to eighty patients, a small Phase II trial might involve fifty to seventy-five patients, and a large one could reach several hundred patients or more. There is no overall FDA-mandated number of patients in each phase. Rather, the goal of each phase is to decide—in light of the patient population, the stage of the disease, the other medicines already approved for the same condition (if any), and so on—whether there is enough evidence to justify going on to the next phase. A new treatment for a tiny "orphan" patient population—where perhaps only a few hundred people around the world suffer from the condition—might be quite small, whereas one for lowering blood pressure in healthy adults could be quite large.

A simple Phase II trial in late-stage cancer patients, for example, might enroll fifty to one hundred patients who had failed all standard treatments, often including surgery, chemotherapy, and radiation. The relevant question to be answered in this patient population could be whether the product can induce remissions in even a few of these very sick patients or perhaps reduce the rates

---

[9] For instance, J. A. DiMasi et al., "Trends in Risks Associated with New Drug Development: Success Rates for Investigational Drugs," *Clinical Pharmacology & Therapeutics* 87, no. 3 (March 2010): 276, show that the Phase I transition probability for small molecules is only 63% while it is 84% for biologics. Similarly, Janice M. Reichert, "Monoclonal Antibodies as Innovative Therapeutics," *Current Pharmaceutical Biotechnology* 9, no. 6 (2008): 427, shows that monoclonal antibodies have Phase I success rates above 80%.

at which the tumors are growing or even show signs of improved survival. Since these patients have exhausted all other therapeutics, a fairly modest showing in Phase II could well encourage a company to move into Phase III trials. At the other end of the spectrum, a Phase II trial of a product in a fairly well-served medical area (heart disease, for example) might require a Phase II trial of hundreds of patients (or more) in at least two "arms," the size of which will generally be determined by the degree of statistical confidence the company (or the FDA) wants to have to commit the resources to larger and more Phase III trials. Patients in one arm would receive the standard treatments now being prescribed for that condition while, in another arm, the patients would receive the new product. Frequently, these multi-arm trials are randomized—meaning that patients coming into the trial are assigned to the various arms randomly to avoid biasing the results by, for example, having all the sickest patients in one of the arms—and may be blind (or double-blind), which calls for the physician not to know whether a patient is receiving the new drug or the standard one. Some trials may even be "placebo controlled," in which one arm of patients gets a placebo (a sugar pill, for example, or another inactive or dummy drug) to see whether the mere act of treatment has an effect on the disease.

Ultimately, Phase II trials are designed to provide the company with enough safety and efficacy information about a product to make a go/no-go decision about Phase III. It is not unusual for Phase II results to come in shades of gray, however. The primary endpoint or other hoped-for outcome has not been achieved, but there is some clear evidence of therapeutic activity. Perhaps, for example, there are some identifiable subgroups of patients who appeared to have a better response to the treatment. Whatever the reason, the mixed signals provided by the results put the company in the position of having to make important decisions under conditions of considerable uncertainty. Should it abandon the product (which, if it is the company's most advanced or only product, is a very hard decision), or instead boldly move into Phase III in one of the positive subgroups without knowing whether the Phase II retrospective subgroup analysis reflects a reproducible outcome or just an artifact of the nature of the particular patients in the Phase II study? To obtain a clearer clinical picture, the company could step back and repeat the Phase II trial with a revised design specifically focused on the subgroup in question. Doing so adds another costly Phase II trial and a lengthy delay on the road to FDA approval. These are hard questions for which there are no clearly correct answers. Moreover, for small loss-making biotechnology companies, the answers may be driven in substantial part by the amount of cash they have and how they think their investors will react to one decision or another. In fact, for venture capital–funded private companies, it may be the investors' risk–reward analysis that is determinative. Even when the company management recommends pressing forward, the venture investors may have better opportunities available in their multicompany portfolios.

For most pharmaceutical products, whether biological or small molecule, Phase III represents the "pivotal trial"—namely, the well-controlled clinical experiment that demonstrates a level of safety and efficacy sufficient to convince the FDA to approve the product and, ultimately, to encourage physicians to prescribe it. Often two well-controlled clinical trials are required. After the completion of Phase II trials, the company typically meets with the relevant staff members at the FDA to discuss how the Phase III trial should be designed. There is a plethora of potential issues for discussion and negotiation, but there is a relatively small handful of issues that revolve around the key question of whether the trial will adequately demonstrate the product's efficacy—in other words, what is the study's endpoint?

In cancer, for example, the clearest endpoint is survival: Do patients who receive the drug live longer than similar patients who are treated with currently approved medicines? Such a trial could last many years since it would take time to enroll and treat all the patients and then wait for a long enough time for a significant portion of the patients in the control group to die. Accordingly, companies often prefer to use alternate endpoints or "surrogate markers" to save time—that is, by measuring a disease-related event (such as tumor shrinkage or control) that is likely to provide the FDA with sufficient evidence to conclude that the patients are receiving a bona fide clinical benefit without waiting for the lengthy conclusion of a "survival" trial. One difficulty with doing so is that with these other endpoints, at least in cancer, successful results using nonsurvival endpoints have not always correlated with enhanced survival. In other words, although the visible traces of malignant tissue may be diminishing, the product may not have, in fact, turned back the course of the disease.

The preceding brief overview of an incredibly complicated process is designed primarily to provide a foundation for a discussion of why creating and developing new medicines is so expensive. Outside the industry itself, perhaps the most frequently discussed and debated issue concerning the drug development process is the cost of completing this lengthy and complex process. Hence, in the remainder of the chapter, we focus on the most recent estimations of those costs as well as the assumptions that are behind those estimations. Before doing so, however, it is worth asking how little companies can carry out all of these complicated steps, especially when most companies have never done it before. Even if they can convince investors to provide the funding, who will do all the work? Here again is a point where the ecosystem—the collection of external resources available to the biotech industry—is essential. Given the complexity involved in designing and executing those clinical studies, not to mention producing adequate quantities of the product under highly regulated conditions that cannot be achieved in a standard research laboratory, it should come as no surprise that only a tiny number of the largest biotech companies have integrated R&D organizations capable of discovering new product candidates and then developing them all the way to the commercial market. The overwhelming majority relies heavily on the expertise and capacity of other companies, such as pharmaceutical company partners (which we discuss in Chapter 5) and contract research organizations (CROs).

As we will discuss further in Chapter 6, the growth of the CRO industry has enabled the phenomenon of the "virtual" biotech companies that sometimes might have just one employee and, in any event, have far fewer employees than they would need to reach FDA approval entirely through their own internal efforts. The expertise of CROs spans the entire spectrum of the drug development process that is depicted in Figure 3.1, from creating initial compounds to performing in vitro, in vivo, and human clinical testing to manufacturing the necessary quantities of the product to applying for FDA approval and even to providing contracted sales and marketing services. The CRO industry has grown rapidly alongside the biotech industry. According to a 2020 industry report, the global CRO market revenue for 2019 was $43 billion, and its revenue is expected to reach $64 billion by 2024, growing at a compound annual growth rate of 8.2%.[10]

With this depth of resources available through CROs, both pharmaceutical and biotech companies can choose to perform virtually any aspect of R&D on an outsourced basis, and that work will often be performed by researchers who have previously been trained and employed by large biotech and pharmaceutical companies. As a result, even the smallest biotech companies have access, if they have the funds, to experienced personnel in all areas of product development. By the same token, pharmaceutical companies, once thought to have an enormous experience-based advantage in drug development, no longer enjoy a practical monopoly in that arena, as large pharmaceutical and small biotech companies increasingly call on the same CROs to do considerable amounts of the R&D work. The key question for the biotech companies,

---

[10] Ben Adams, "CRO Market to Recover, Using 'Hybrid trials,' with Revenue Hitting $64B by 2024: Report," *FiercePharma*, September 8, 2020, https://www.fiercebiotech.com/cro/cro-market-to-reco ver-using-hybrid-trials-revenue-hitting-64b-by-2024-report.

then, is whether they can access enough funding to keep their products going through the process.

The Two-and-a-Half-Billion-Dollar Drugs

This brings us back to the question of how much all of these R&D activities cost. The best-known studies, which have been done primarily at the Tufts Center for the Study of Drug Development, have estimated the cost of developing a new drug at $2.56 billion (in 2013 dollars). The authors of the Tufts Study tracked data from the development of 106 randomly selected drugs developed by ten different pharmaceutical companies over a period of about twenty years to determine with as much precision as possible how much it costs to develop a new drug. The final number, $2.56 billion, is daunting, to say the least. Simple multiplication would suggest that today's biotechnology industry of several thousand companies would need to convince extremely patient investors to bestow upon them nearly *ten trillion* dollars to allow each company to develop just one product. Even during those occasional times of plenty when Wall Street becomes infatuated with biotech's potential for medical miracles, such a figure seems beyond the realm of possibility. To see how the biotech industry might be able to overcome these severe financial hurdles, it is important to parse through the process of drug development and to analyze the financial methodology of the Tufts Study. Will it really cost each biotech company $2.5 billion to develop a new medicine? Fortunately, the answer is no.

The first thing we can do is take the total number derived by the Tufts researchers—$2.56 billion—and cut it almost in half. Around 45% ($1.16 billion) of the full costs are attributable to the lost economic opportunities created by devoting a great deal of money to the long process of drug development instead of investing the same amount in something else, such as income-producing securities. That is, if all the money that is spent in drug development were instead invested in stocks and bonds, the investor would have a lot more money at the end of a dozen years than at the beginning because of the financial miracle of compound interest. (For example, money invested at a 10% compounding interest rate will double just about every five years.) So, in the Tufts analysis, the researchers focused not just on out-of-pocket expenditures but have applied the well-known economic analysis technique that says that to evaluate an investment in something like drug development, it is important to compare that investment with what would have happened if the investor had chosen to invest in something else.

After all of these "opportunity costs" are factored in, the actual money spent in the drug development process is estimated at $1.4 billion. These amounts include the salaries and benefits of the pharmaceutical researchers, the costs of clinical testing, and the myriad other expenditures associated with the research and development process (which the Tufts Study refers to as "out-of-pocket" costs). For the purposes of sound economic analysis, it is certainly reasonable to include the other $1.16 billion of lost opportunity costs, but for biotech companies there are no alternative investment options. In fact, investors have invested in biotech companies on the assumption that they will be single-mindedly devoted to the creation of new medicines. And so, to understand the cash needs of the biotech industry, we need to focus on the out-of-pocket costs.

While the lost opportunity costs essentially double the overall costs of drug development, another factor plays an even more dramatic role. The single largest cost associated with the development of any successful pharmaceutical product is the cost of all the unsuccessful drugs that had to be discovered, developed, and tested to find the one that finally works. There are a variety of studies that have attempted to calculate the number of compounds synthesized at the beginning of the research process to yield one successful drug at the FDA approval stage, and estimates invariably run in the thousands. Even at the relatively advanced stage of entering clinical development, which only happens after extensive laboratory and animal testing, the failures far outnumber the ultimate successes. In looking at drugs entering clinical development, the Tufts Study researchers concluded that 11.83% of the product candidates included in their analysis had reached (or would reach) FDA approval. That is, nearly 88% of the new medicines reaching the relatively advanced stage of human testing eventually fall by the wayside. (This 88% failure rate is probably unduly optimistic. The Tufts Study focuses on products entering clinical trials in the United States, and therefore the success rates are influenced by the fact that some of the products entering U.S. clinical trials have already been through significant levels of clinical testing in other countries, sometimes all the way to the point of regulatory approval.) There is attrition at every step: not all of the products are funded through Phase III trials, and over a third fail during Phase I, but there is little question that the expense of drugs that fail far exceeds the cost of those that succeed.

The median *clinical* cost of drug development for one drug to complete the entire program (that is, Phase I–III trials plus the long-term animal studies required for FDA approval) is approximately $262 million, and the corresponding mean cost is $339 million, according to the Tufts Study.[11] Those

---

[11] The mean is the mathematical average of a set of numbers, and the "median is the middle number in a sorted, ascending (or descending), list of numbers . . ." Akhilesh Ganti, "Median," *Investopedia*, August 22,

numbers are calculated by summing up the median (mean) costs of each phase of clinical trials. It should also be noted that those numbers do not represent the actual out-of-pocket costs of any particular product, but the median and mean costs of all products in the study; thus, some products could be much more expensive to develop while others might be developed for quite a bit less.[12]

Because a drug can fail at any phase during the development process, the study also provides the expected cost of a drug undergoing clinical trials. This is the weighted average of estimated mean phase costs, where the weights are based on the estimated probabilities that a drug will enter a given phase. According to the study, once those failure rates are accounted for, the expected cost for each drug entering clinical trials is $114.2 million. Moreover, given that only 11.83% of drugs entering clinical trials eventually succeeds in gaining FDA approval, the study concludes that the out-of-pocket clinical cost per approved drug is $965 million. The difference between the out-of-pocket cost and the total expected cost per approved drug highlights the colossal cost of failures in drug development.

Having pegged the total clinical costs at $965 million (successes plus failures), the Tufts Study authors then calculate how much had to be spent on discovery and preclinical research to generate enough clinical candidates to ultimately lead to one FDA-approved drug. That number is $430 million, a figure they derive by analyzing research costs during the roughly five-year "lag time" from the creation of a new compound until its entry into clinical testing, including the cost of compounds that fail along the way. Adding the preclinical and clinical costs together yields the $1.4 billion total out-of-pocket expenses per approved drug cited earlier.

Not only are the costs of drug development enormous, but they also appear to be growing at an alarming rate. As noted at the beginning of this chapter, according to the Tufts Study, the total capitalized development costs grew from $1.04 billion in 2003 to $2.56 billion in 2013 (expressed in each case in

2019, https://www.investopedia.com/terms/m/median.asp. A common example illustrating the difference between mean and median involves Bill Gates walking into a bar just as the person who previously had the highest net worth leaves. The result is that the mean goes up dramatically, but the median stays the same. See Dan Ma, "When Bill Gates Walks into a Bar," *Introductory Statistics*, September 4, 2011, https://introductorystats.wordpress.com/2011/09/04/when-bill-gates-walks-into-a-bar/.

[12] One 2013 analysis of the per-patient clinical trial costs showed a low of $16,500 in infectious diseases to a high of $59,500 in oncology, with an overall average of $36,500. The study estimated that, in 2013, 1,148,340 patients were included in industry-sponsored clinical trials. See Battelle Technology Partnership Practice, *Biopharmaceutical Industry-Sponsored Clinical Trials: Impact on State Economies*, March 2015, 5, 7, accessed online, http://phrma-docs.phrma.org/sites/default/files/pdf/biopharmaceutical-industry-sponsored-clinical-trials-impact-on-state-economies.pdf.

2000 dollars, meaning that the overall effects of inflation have been taken into account). The sharp increase is primarily driven by the increase in the out-of-pocket costs, rather than cost-of-capital ones. Specifically, the out-of-pocket costs have grown from $0.52 billion in 2003 to $1.4 billion in 2013 (both figures in 2013 dollars). As the authors explain, almost half of this 166% increase in out-of-pocket costs can be attributed to the significantly higher failure rates of clinical trials observed in the most recent study compared with the previous one. According to the 2003 study, the odds of ultimately obtaining FDA approval of a drug entering clinical trials was almost 20%, but the corresponding probability for 2013 was just under 12%, a figure consistent with other industry studies.[13]

Researchers and industry analysts have offered a number of reasons that could explain this drop: the "better than the Beatles" problem, where new drugs are designed to be not only as good as but also better than the best current standard of treatment; the focus on more challenging targets and diseases; the increasing complexity of clinical trials and the hurdles for regulatory approval as well as reimbursement pressures (for example, demonstrating benefit with respect to comparator drugs rather than placebos); etc.[14] While it is beyond the scope of our analysis to discuss all of the specific drivers behind those high failure rates, it is important to highlight that it is precisely those failures, rather than the few successes, that primarily determine the overall costs of the entire drug development process.

As a case in point, consider the controversy generated in light of the study published by Vinay Prasad and Sham Mailankody in 2017 that estimated the cost of developing a new cancer drug not at over $2.5 billion but at $648 million.[15] What could explain such a large difference between that study and the Tufts estimate? As noted in several follow-up articles,[16] a fundamental methodological flaw in the Prasad and Mailankody study is the so-called survival bias. The latter occurs when the analysis includes only subjects that survived

[13] See Hay et al., "Clinical Development," 40–51.

[14] See, for example, Fabio Pammolli, Laura Magazzini, and Massimo Riccaboni, "The Productivity Crisis in Pharmaceutical R&D," *Nature Reviews Drug Discovery* 10, no. 6 (June 2011): 428–38; and Jack W. Scannell et al., "Diagnosing the Decline in Pharmaceutical R&D Efficiency," *Nature Reviews Drug Discovery* 11, no. 3 (March 2012): 191–200.

[15] See Vinay Prasad and Sham Mailankody, "Research and Development Spending to Bring a Single Cancer Drug to Market and Revenues after Approval," *JAMA Internal Medicine* 177, no. 11 (November 2017): 1569.

[16] See, for example, Peter J. Pitts, "A Flawed Study Depicts Drug Companies as Profiteers," *Wall Street Journal*, October 9, 2017, https://www.wsj.com/articles/a-flawed-study-depicts-drug-companies-as-profiteers-1507590936; and Matthew Herper, "The Cost of Developing Drugs Is Insane. That Paper That Says Otherwise Is Insanely Bad," *Forbes*, October 16, 2017, https://www.forbes.com/sites/matthewherper/2017/10/16/the-cost-of-developing-drugs-is-insane-a-paper-that-argued-otherwise-was-insanely-bad/#ac88ceb2d459.

or succeeded, ignoring the ones that for whatever reason did not. To get a sense of how such omissions can affect the results, consider, for example, estimating the return of buying a lottery ticket with a sample that only included lottery ticket winners. In the case of Prasad and Mailankody, their study only includes data from ten companies that had *successfully* developed and marketed one drug each. Given the relatively small size of those ten companies (the median R&D pipeline size was 3.5 drugs), these companies won the biotech equivalent of the lottery. As we discuss at greater length in Chapter 7, the biotech industry has initiated about 40,000 different R&D programs over the last two decades. Only 245 have been approved by the FDA. (See Figure 3.2 for FDA approvals by year.) The Prasad and Mailankody study indicates that, for ten very fortunate companies, they were able to beat the odds and obtain a cancer approval for (only) $648 million. But the actual cost of the drug development process will only begin to resemble $648 million when one out of every 3.5 cancer product development programs ends in success. According to the most recent study of success rates, the likelihood of success of a clinical stage cancer product is less than one out of every sixteen. Because the Prasad and Mailankody analysis fails to account for all the hundreds or thousands of companies that dedicated resources to cancer drug candidates, but, for whatever reason, were not able to reach FDA approval, the $648 million estimate does not represent what it costs for either the biotech or pharmaceutical industry to generate a new product.

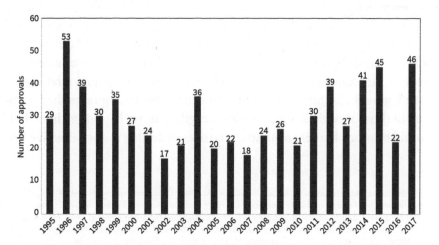

Figure 3.2 Number of new molecular entities (NMEs) and biologics license applications (BLAs) approved by the FDA
Source: www.fda.gov.

Conclusion

In this chapter, we have offered a very brief overview of a drug development process that would require several volumes to describe in any detail, and, therefore, we have listed a number of such volumes in the notes.[17] The point of the overview of the various stages of development is to provide a foundation for the rest of the chapter's focus on the costs of creating a new FDA-approved medicine. It also offers a view of the common vocabulary shared by all of the various parties involved in the biotech industry's breakthrough-to-blockbuster process. When an IND will be filed, or when Phase III will commence, and so on, are the kinds of questions routinely raised by investors, potential partners, reporters, and others in the industry. The answers offer a rough guide to where the company fits in the universe of biotech companies and how close (or far) it is from having an FDA-approved product.

We have also discussed the most recent figures regarding the cost of drug development. Reliable estimates about the cost of drug development are critical, not just for the companies developing those drugs and their investors but also for policymakers. We have centered most of our discussion on the Tufts Study, not only because it is the one most often cited but also because it provides a detailed description of its methodology, underlying assumptions, and comparisons with other studies aimed at estimating the cost of drug development. Given the wide range of different drugs, combined with the complexity of the various clinical trial pathways involved in each indication, getting a single-point precise estimate about the average cost per FDA approval is far from trivial. Yet if we follow the Tufts Study across the past three decades, the trend is clear: the cost per approved drug was $473 million in 1991,[18] $1.04 billion in 2003, and $2.56 billion in 2013 (all figures in 2013 dollars). Those constantly growing costs seem to have made no difference in the number of FDA approvals, which have remained relatively flat (see Figure 3.2). Thus, it is possible to debate the assumptions that go into any of those models at much greater length than we have done here, but it would be much harder to question the overall decline in cost-effective R&D

---

[17] See, for example, Donald R. Kirsch, "Therapeutic Drug Development and Human Clinical Trials," in *Biotechnology Entrepreneurship: Leading, Managing and Commercializing Innovative Technologies*, ed. Craig Shimasaki, 2nd ed. (London: Academic Press, 2020), 315–50; Sarwar Beg et al., eds., *Pharmaceutical Drug Product Development and Process Optimization: Effective Use of Quality by Design* (Palm Bay, FL: Apple Academic Press, 2020); Lawrence M. Friedman et al., *Fundamentals of Clinical Trials*, 5th ed. (New York: Springer, 2015); and Daan J. A. Crommelin, Robert D. Sindelar, and Bernd Meibohm, eds., *Pharmaceutical Biotechnology: Fundamentals and Applications*, 5th ed. (New York: Springer, 2019).

[18] See Joseph A. DiMasi et al., "Cost of Innovation in the Pharmaceutical Industry," *Journal of Health Economics* 10, no. 2 (July 1991): 107–42.

productivity at the industry level over the past few years.[19] This highlights the importance of identifying effective R&D strategies, a topic that we will discuss in Chapter 7.

[19] See, for example, Stacy Lawrence, "R&D Growth Stalls," *Acumen Journal of Sciences* 1, no. 2 (August/ September 2003): 22; Robert F. Service, "Surviving the Blockbuster Syndrome," *Science* 303, no. 5665 (March 2004): 1796–99, noting on p. 1797 that a "staggering 99.9% of compounds wash out of the development pipeline"; Bernard Munos, "Lessons from 60 Years of Pharmaceutical Innovation," *Nature Reviews Drug Discovery* 8, no. 12 (December 2009): 959–68; Pammolli, Magazzini, and Riccaboni, "Productivity Crisis," 428–38; and Scannell et al., "Diagnosing the Decline," 191.

4

The Companies

At any particular time during the twenty-first century, there have been over 5,000 biotech companies scattered throughout the world, with the largest concentration of them in the United States, where there are typically over 2,000 companies. This impressive number can be compared with 100 or so companies that make up the traditional pharmaceutical industry, of which perhaps only about 50 account for the lion's share of pharmaceutical sales.

Some biotech companies are constructed by venture capitalists specifically for the purpose of being acquired by large pharmaceutical companies, while many others have set out to become one of those top handful of pharmaceutical giants. These plans for long-term success and independence gain inspiration and enthusiasm from Amgen, which has enjoyed stunning success. Incorporated in 1980, Amgen completed its IPO in 1983 at a valuation of $190 million.[1] Eleven years later, Amgen had successfully launched two blockbuster products and saw its market value surge to $10 billion. By 2017, still largely on the strength of those two products, Amgen's market value reached $132 billion, placing it firmly within the top tier of pharmaceutical companies, with a market capitalization higher than traditional powerhouses such as Bristol Myers Squibb. Amgen's blockbuster-based success has followed a recognizable pattern within the pharmaceutical industry. It is not uncommon for biopharmaceutical firms to achieve their success through the launch of one or two extremely valuable products. That an important new drug can reach several billion dollars per year in sales within a few short years after market launch is one of the reasons investors become excited about the industry.

Until a biotech company achieves Amgen-like commercial success, which will allow it to fund its own R&D efforts, it needs to raise cash from investors, which is why one veteran investment banker, Stelios Papadopoulos, said to a meeting of biotech executives one of us attended, "Until you get a product on the market, all you've got to sell is stock." No matter how exciting the

---

[1] See Stelios Papadopoulos, "Evolving Paradigms in Biotech IPO Valuations," *Nature Biotechnology* 19, no. 6 (June 2001): BE 18–19.

company's scientific work may be, the money needed for those research and development efforts has been as important to the company's overall success as the molecules being developed. Since developing new pharmaceutical products can cost hundreds of millions, or even billions, of dollars, the search for a continuing supply of cash resources has been an essential component of the biotech industry. In this chapter, we will look at the primary ways that these two critical biotech industry components—molecules and money—manage to get together.

Biotech's Academic Origins

Biotech companies are the offspring of a marriage of convenience between science and money. These two essential components are jointly invested in the pursuit of profits and promising medicines. In some cases, investors seek out biomedical technologies. In others, scientists with potential medical products or technology set out in search of funding for their work. Rarely, if ever, are these the scientists of entrepreneurial lore who tinker in oil-stained garages and musty basements stacked to the ceiling with beakers, Bunsen burners, and pipettes. Modern biology requires stunningly expensive equipment and sterile conditions in which impurities are measured in parts per billion. As a result, the largest number of biomedical scientists outside the industry work not in garages, but as faculty members in research universities around the world, where government grants, foundation support, and charitable contributions make their expensive scientific endeavors possible. In many countries, including the United States, these professors receive their salaries from the university, but the funding for those salaries—and for the costs of running their laboratories—is heavily dependent on grants from government organizations such as the National Institutes of Health. This funding arrangement makes the first key question for anyone planning to launch an entrepreneurial biotech company based upon research in an academic laboratory: How does the ownership of that government grant-funded university research end up in the hands of a new, for-profit company? In the United States, where the biotech industry was launched and where the greatest concentration of companies can still be found, the answer can be found in an Act of Congress.

Until 1980, any technology developed in the laboratories of U.S. government-funded faculty members would belong to the government. The federal government was not especially well equipped to take advantage of the vast amount of technology flowing from its funded research: by 1978, the government had licensed for commercial use less than 5% of the

25,000–30,000 patents it owned.[2] In 1980, Congress passed the Bayh–Dole Act, which was designed to provide a more efficient technology transfer platform by allowing universities to patent discoveries made at their institutions.[3] Following the Bayh–Dole Act, all discoveries made by university faculty members would belong to that academic institution, which would have the power to decide which inventions to patent and how to find a company to take on the responsibility for the commercial development of any products that might be based on the technology. The universities would get to keep the proceeds of any licensing arrangements, and they established internal policies providing for how such funds would be divvied up. A typical approach is Stanford's, which gives 15% to the technology transfer office, with the remainder divided into thirds: one to the faculty member who made the discovery, one to the department, and one to the faculty member's school.[4]

The concept behind the Bayh–Dole Act, which clearly put universities and researchers into the stream of commerce, has not always been universally praised. Newspaper columnist Ellen Goodman argued, for example, that linking "science and business, nonprofit and profit" combines "altruism and chumphood," and even technology transfer experts have pointed out that "of almost 21,000 licenses [from universities to companies] active in 2000, less than 1% generated income in excess of $1 million."[5] Yet there is no question that the licenses that have been productive have made significant contributions to the economy and to the universities themselves. According to Washington attorney Michael Remington, as of 2002, when Goodman issued her critique, "sales of goods developed from products transferred from university research centers produced a whopping $42 billion in revenues, and US universities, research institutes, and hospitals recouped almost $1.2 billion in gross income."[6] These products range from the sports drink Gatorade to numerous modern medicines for cancer and other life-threatening conditions. A few years later,

---

[2] See Bradley Graham, "Patent Bill Seeks Shift to Bolster Innovation," *Washington Post*, April 8, 1979, https://www.washingtonpost.com/archive/business/1979/04/08/patent-bill-seeks-shift-to-bolster-innovation/db14f277-ec0e-4ca5-9aeb-ce2cad86e25b/.

[3] See Patent and Trademark Law Amendment (Bayh–Dole) Act, Pub. L. No. 96-517, 94 Stat. 3015 (codified as amended at 35 U.S.C. §§ 200–12) (1980). For an extended discussion of the Bayh–Dole Act and its role in university technology transfer in the United States, see David C. Mowery et al., *Ivory Tower and Industrial Innovation: University-Industry Technology Transfer before and after the Bayh-Dole Act* (Stanford, CA: Stanford University Press, 2015).

[4] See "9.1 Inventions, Patents, and Licensing," Research Policy Handbook, Stanford University, last modified June 19, 2013, https://doresearch.stanford.edu/policies/research-policy-handbook/intellectual-property/inventions-patents-and-licensing.

[5] Ellen Goodman, "Medicine Needs More 'Chumps,'" *Boston Globe*, March 1, 2001, A15.

[6] Michael J. Remington, "A Board's Primer on Technology Transfer," *Trusteeship* 10, no. 6 (November/December 2002): 17. There are periodic efforts to revise the Bayh–Dole Act. One call for such an effort is Arti K. Rai and Rebecca S. Eisenberg, "Bayh-Dole Reform and the Progress of Biomedicine," *American Scientist* 91, no. 1 (January/February 2003): 52–59.

Senator Birch Bayh, coauthor of the Bayh–Dole Act, reported the effects of university patent licensing on the U.S. economy between 1996 and 2010:[7]

- up to a $388 billion impact on the U.S. gross domestic product;
- up to a $836 billion impact on U.S. gross industrial output; and
- up to 3 million new jobs created in the U.S. from university inventions.

Data compiled by the Association of University Technology Managers (AUTM) show that universities and research institutes in the United States granted 6,372 new licenses (and option agreements) in 2012, and 705 new start-up companies were formed that year specifically to develop commercial applications from academic research.[8] According to the latest estimates, over the past four decades, the Bayh–Dole Act has bolstered U.S. economic output by $1.3 trillion, supported 4.2 million jobs, and helped lead to more than 11,000 start-up companies.[9] Although these licenses and companies cover all industries and thus extend well beyond the life sciences, a significant percentage relates to biomedical technologies. AUTM no longer asks its members to provide data by industry segment. But its surveys in the late 1990s show that the "life sciences" consistently made up nearly 70% of all licensing and nearly 85% of licensing income for U.S. universities, hospitals, and research institutions.[10] A 2015 analysis of the top U.S. universities (ranked by the total number of licensing agreements) found that life sciences accounted for more than 70% of the total number of licenses and option agreements and that those licenses accounted for over 93% of their gross technology transfer revenues.[11]

[7] See Association of University Technology Managers, *The AUTM Briefing Book: 2015*, 5, accessed online, https://www.cshl.edu/wp-content/uploads/2017/12/AUTM-Briefing-Book-2015.pdf. See also Loris Pressman et al., *The Economic Contribution of University/Nonprofit Inventions in the United States: 1996–2013* (Washington, DC: Biotechnology Industry Organization, 2015), accessed online, https://www.bio.org/sites/default/files/legacy/bioorg/docs/BIO_2015_Update_of_I-O_Eco_Imp.pdf.

[8] See Association of University Technology Managers, *2012 AUTM U.S. Licensing Activity Appendix* (Deerfield, IL: AUTM, 2013).

[9] See Bayh–Dole 40, accessed September 1, 2020, https://bayhdole40.org.

[10] In 1996, life sciences made up 68% of licensing and 84% of licensing income for U.S. universities, hospitals, and research institutions: Association of University Technology Managers, *AUTM Licensing Survey: FY 1996*, ed. Daniel E. Massing (Norwalk, CT: AUTM, 1997), 9–10, figs. 4–5. In 1997, life sciences made up 70% of their licensing and 85% of their licensing income: Association of University Technology Managers, *AUTM Licensing Survey: FY 1997*, ed. Daniel E. Massing (Norwalk, CT: AUTM, 1998), 9, figs. 4–5. Drivas, Lei, and Wright observe that 75% of invention disclosures in the University of California system over five years were in the life sciences: Kyriakos Drivas, Zhen Lei, and Brian Wright, *A Preliminary View of University of California Data on Disclosures, Licensing and Patenting* (Washington, DC: National Academy of Sciences, 2009), 1, accessed online, https://sites.nationalacademies.org/cs/groups/pgasite/documents/webpage/pga_058895.pdf.

[11] See Brady Huggett, "Top US Universities, Institutes for Life Sciences in 2015," *Nature Biotechnology* 35, no. 3 (March 2017): 203.

In the words of a report to Congress by the National Institutes of Health, technology transfer from universities to companies in the life sciences has "yielded a dramatic return to the taxpayer through the discovery of new technologies that extend life and improve the quality of life through the development of products that might not [otherwise] be available," and it has "buttress[ed] the biomedical research enterprise that has made the United States the world leader in this field."[12] It is little wonder that a desire to continue the trend of "more than 6,000 new U.S. companies formed from university invention . . . is a key ingredient in virtually every state economic development plan."[13]

The interactions between academia and the biotech industry do not end with the initial technology transfer and company launch. A survey published in the *New England Journal of Medicine* indicates that nearly all biotech companies supported academic research,[14] and about one-third of all academic biomedical researchers had financial ties to industry at that time.[15] Another analysis shows that more than half of the academic life sciences researchers surveyed had "some form of relationship with industry."[16] With cutting-edge research taking place in both academia and industry, these collaborations have the potential to be mutually beneficial. In fact, these relationships have regularly been the means by which academic research has been developed into FDA-approved medicines.[17] One study reported that research in the life

[12] U.S. Department of Health and Human Services, National Institutes of Health, *National Institutes of Health Response to the Conference Report Request for a Plan to Ensure Taxpayers' Interests Are Protected* (Washington, DC, 2001), 19. Not surprisingly, AUTM's tenth anniversary report celebrated "the wisdom of the U.S. Congress in the enactment of the Bayh–Dole Act of 1980": Association of University Technology Managers, "A Message from the President," in *AUTM Licensing Survey: FY 2000*, ed. Lori Pressman (Northbrook, IL: AUTM, 2002). To buttress AUTM's claims to the dual success of the Bayh–Dole Act and AUTM's 2,800 technology-transfer professional members, its president cited a report by the National Institutes of Health to Congress in 2001, which reads: "Current practices in technology transfer have yielded a dramatic return to the taxpayer through the discovery of new technologies that extend life and improve the quality of life through the development of products that, without the successful public-private relationship, might not be available. The transfer of federally funded technology has also resulted in financial returns from licensing activity; and such funds are used to buttress the biomedical research enterprise that has made the U.S. the world leader in this field" (National Institutes of Health, *Response*, 19).

[13] Quoted in Donald Drakeman, *Why We Need the Humanities: Life Science, Law and the Common Good* (New York: Palgrave Macmillan, 2016), 19.

[14] David Blumenthal et al., "Relationships between Academic Institutions and Industry in the Life Sciences—An Industry Survey," *New England Journal of Medicine* 334, no. 6 (February 1996): 368–73. For a broader discussion of these relationships, see David Blumenthal, "Academic–Industrial Relationships in the Life Sciences," *New England Journal of Medicine* 349, no. 25 (December 2003): 2452–59.

[15] See the various studies cited in Justin E. Bekelman, Yan Li, and Cary P. Gross, "Scope and Impact of Financial Conflicts of Interest in Biomedical Research: A Systematic Review," *Journal of the American Medical Association* 289, no. 4 (January 2003): 454–65.

[16] Darren E. Zinner et al., "Participation of Academic Scientists in Relationships with Industry," *Health Affairs* 28, no. 6 (November 2009): 1814–25. A full 53% of those surveyed indicated such a relationship.

[17] See Robert Kneller, "The Importance of New Companies for Drug Discovery: Origins of a Decade of New Drugs," *Nature Reviews Drug Discovery* 9, no. 11 (November 2010): 869, which attributes 24% of new drugs to universities. For reasons discussed later, we believe that universities' contributions may actually be considerably higher.

sciences has resulted in at least "153 new FDA-approved drugs, vaccines, or new indications for existing drugs" between 1970 and 2009, with additional new medicines appearing every year.[18] Importantly, these discoveries originated from a wide range of universities and research institutions, rather than just a short list of consistent "outperformers."[19]

The Bayh–Dole Act was certainly one of the critical factors that ignited an explosion of biotechnology start-ups in the 1980s, but there were other important components of the biomedical revolution as well. At roughly the same time, prototypical biotech companies such as Genentech were able to complete wildly successful IPOs on the strength of little more than public enthusiasm for their interesting technology.[20] That event showed venture capitalists that it was possible to make money—potentially, a lot of money—long before new biological discoveries were turned into profitable products. The IPO opened up a public market in which the venture capitalists could sell the stock they had acquired when they invested in the start-up company.

Genentech's successful IPO in 1980 not only opened the door for several generations of biotechnology companies to complete IPOs at early stages of development, but its high-flying IPO also sent venture capitalists scurrying around the scientific world to find scientists or scientific discoveries that might serve as the foundation for new Genentech-like biotech companies. And so, as a result of the Bayh–Dole Act, scientists were looking for industrial partners to help turn their discoveries into new medicines. At the same time, investors were looking for scientists with good ideas and patentable technology. Helping catalyze these interactions between research science and biomedical investors was the War on Cancer declared by President Nixon in 1971, which poured vast amounts of federal funding into research. Cancer has remained a governmental priority ever since, and there was a total of $7 billion just for cancer research at the National Institutes of Health in the federal budget over forty years later, plus billions more devoted to cancer research via charitable grants and pharmaceutical R&D spending.[21]

[18] Ashley J. Stevens et al., "The Role of Public-Sector Research in the Discovery of Drugs and Vaccines," *The New England Journal of Medicine* 364, no. 6 (February 2011): 535.

[19] Stevens et al., "Public-Sector Research," Supplementary Appendix, accessed online, https://www.nejm.org/doi/suppl/10.1056/NEJMsa1008268/suppl_file/nejmsa1008268_appendix.pdf.

[20] See Sally Smith Hughes, *Genentech: The Beginnings of Biotech* (Chicago: University of Chicago Press, 2011), chap. 6. "Genentech issued a million shares of stock in its Initial Public Offering (IPO), at an opening price of $35 each. Within an hour, the stock—for those who got their hands on it—leapt to $88 a share, closing at $71.25. *The Wall Street Journal* called it 'one of the most spectacular market debuts in recent history'"; see Laura Fraser, "Genentech Goes Public," Genentech, April 28, 2016, https://www.gene.com/stories/genentech-goes-public.

[21] Calculated from National Institutes of Health, *Estimates of Funding for Various Research, Condition, and Disease Categories (RCDC)*, March 7, 2014, https://report.nih.gov/categorical_spending.aspx. See also Clifton Leaf, "Why We're Losing the War on Cancer (and How to Win It)," *Fortune*, March 22, 2004, https://fortune.com/2004/03/22/cancer-medicines-drugs-health/. The article notes that in total, "Americans

After Bayh–Dole and the Genentech IPO, academic scientists and venture capitalists, who are neither natural allies nor philosophical compatriots, began spending time together. The venture firms would hire leading academic scientists to serve as consultants and advisors in the hopes that these experts might point out the next most exciting technological breakthrough. At the same time, universities began hiring MBAs and JDs with industry experience to staff the technology transfer offices that would broker the deals between the scientists and either venture capital firms or young biotech companies. By 2005, for example, MIT's office of numerous full-time licensing professionals (plus a sizable patent and financial staff) had "over 650 active technology licenses in house, of which over 100 are extant startup companies."[22] Recently, many universities have further stepped up their focus on spurring entrepreneurial activity by significantly enhancing the traditional role of the technology transfer office. For example, in 2014 UCLA created a not-for-profit company, Westwood Technology Transfer, to focus on "protecting and optimizing" UCLA's discoveries and inventions.[23] During the same year, the University of California launched a $250 million fund to be invested in startups emerging from the University of California system.

At the same time, these kinds of links between academia and industry have led to concerns over conflicts of interest and the overall direction of academic research. Specifically, some scholars have argued that the growing "privatization and erosion of the scientific commons" slows down scientific progress as it leads to delays in sharing the knowledge and divert faculty from basic research.[24] Yet a number of other research studies have found no support for those hypotheses.[25] These conflicts of interest and conflict of commitment questions have nevertheless been intensely debated for the past several decades, and the universities have created a variety of policies to deal with them. Imperial Innovations, for example, has described the balancing effort involved in forming a new "spin-out" company from academic research:

have spent, through taxes, donations, and private R&D, close to $200 billion, in inflation-adjusted dollars, since 1971."

[22] Lita Nelsen, "Technology Licensing Office," Reports to the President, March 18, 2015, http://web.mit.edu/annualreports/pres05/03.18.pdf.

[23] See Brady Huggett, "Reinventing Tech Transfer: US University Technology Transfer Offices Are Adopting New Models in Search of Increased Return on Research Investment," *Nature Biotechnology* 32, no. 12 (December 2014): 1185.

[24] See, for example, Richard R. Nelson, "The Market Economy, and the Scientific Commons," *Research Policy* 33, no. 3 (April 2004): 455–71.

[25] See, for example, Jerry G. Thursby and Marie C. Thursby, "Who Is Selling the Ivory Tower? Sources of Growth in University Licensing," *Management Science* 48, no. 1 (January 2002): 90–104; and John P. Walsh, Wesley M. Cohen, and Charlene Cho, "Where Excludability Matters: Material versus Intellectual Property in Academic Biomedical Research," *Research Policy* 36, no. 8 (October 2007): 1184–203.

Involvement in the spinout can also pull you in many different directions and can result in possible conflicts of interest, which must be appropriately managed. Please see Imperial's Conflict of Interest Policy. . . . Situations may arise in which the academic founders, other directors and managers of spinout companies find themselves faced with a conflict of interest that arises after formation. It is not enough to clear conflicts only before formation—you must review the situation constantly. This should be declared to the company board as soon as that person is aware of the potential conflict.[26]

Whether these kinds of policies are fully effective in practice will likely continue to be a matter of discussion and debate both on campus and off. Yet the potential for conflicts of interest may be inevitable in the post-Bayh–Dole academic world because the only way for the universities to fulfill their mandate under the act "to promote the commercialization and public availability of inventions made in the United States" is to become something of a business partner with industry. For the purpose of our analysis of the industry, it is sufficient to note that academia has grown sufficiently comfortable with its role in biomedical innovation, international patenting, and active licensing that many of the major advances in the biotech industry have their roots in university research and that the results of that licensing continue to provide billions of dollars to the universities.[27]

In the context of these generally glowing reports about the key role played by technology transfer both on campuses and in the biopharmaceutical ecosystem, there has been little empirical research on their performance. Therefore, there has been a lack of understanding of the factors that are likely to affect the chances of success and failure of university-licensed start-ups. A notable exception is the recent work of Paul Godfrey, Gove Allen, and David Benson, who have tracked the "life history" of academic spin-off companies, including key events such as fundraising, acquisition, IPO, and bankruptcy/dissolution.[28] They point out that focusing only on granting licenses or starting companies tells only the beginning of the story.[29] As in virtually every field, many start-ups fall by the wayside. The authors' results show that approximately half of them either no longer exist (for example, cases of

---

[26] Imperial College London, *A Founder's Guide to Spinouts—Your Guide to Starting a Spinout Company at Imperial College London* (London: Imperial Innovations, 2017), 18, 130, accessed online, https://www.imperial.tech/media/uploads/files/A_Founders_Guide_to_Spinouts_-_Second_Edition_WEB.pdf.

[27] Association of University Technology Managers, *2012 AUTM U.S. Licensing Survey* (Deerfield, IL: AUTM, 2013).

[28] Paul C. Godfrey, Gove N. Allen, and David Benson, "The Biotech Living and the Walking Dead," *Nature Biotechnology* 38, no. 2 (January 2020): 132–41.

[29] See Godfrey, Allen, and Benson, "Walking Dead," 132.

bankruptcy or dissolution) or end up being the so-called walking dead: companies kept alive on paper but without any significant contribution to employment, patents, or commercial products.[30] Ultimately, of the 498 companies that constitute their final data set, only 66 companies (13.3%) were eventually acquired (an event often leading to a highly profitable "exit"), and only 51 (10.2%) went through an IPO (an event that led to more funding, but not necessarily resulting in profitability).[31]

As these many post-Bayh–Dole activities led to entrepreneurial companies and job creation in the United States, other countries set out to emulate those technology transfer activities. Whereas in 2001, two decades after Bayh–Dole, North American academic institutions generated over a billion dollars in licensing revenues, with Columbia University alone receiving $130 million, the leading U.K. institution that year, the University of Edinburgh, received about $5 million. The *Times Higher Education Supplement* quoted a U.K. business school researcher as saying that "for each pound spent on research, the UK outperforms the US at making discoveries but it does not make [them] pay."[32] Then, by 2003, London's Imperial College was spawning spin-off companies at the rate of one per month.[33] A little over a decade later, in 2016, Imperial Innovations (the technology transfer office of Imperial College) invested nearly £70 million across thirty-three portfolio companies,[34] while Cambridge Enterprise (the technology transfer office of the University of Cambridge) reported £16.9 million in operating income raised from licensing and consulting activities and 126 licenses signed in a single year.[35]

These successes were part of a broader effort by the United Kingdom to encourage technology transfer, especially in the life sciences. As Lord Sainsbury, the U.K. Science and Innovation Minister, noted back in 2002, "Fifteen of the world's top 75 medicines were discovered and developed in Britain," and his goal was to stay at the forefront of medical innovation through a series of nineteen seed funds involving fifty-seven different academic institutions.[36]

---

[30] See Godfrey, Allen, and Benson, "Walking Dead," 134.

[31] See Godfrey, Allen, and Benson, "Walking Dead," 132. Another key finding of their work is that while the role of the major U.S. life science clusters is indisputable (their sample is focused on U.S. university-licensed companies), what matters even more is matching the needs of a particular start-up to the resources offered by a given cluster. See Godfrey, Allen, and Benson, "Walking Dead," 137.

[32] Caroline Davis, "UK Poor Cousins to US Tech Transfers," *Times Higher Education*, June 6, 2003, 60.

[33] See Susan Searle, Brian Graves, and Chris Towler, "Commercializing Biotechnology in the UK," *Bioentrepreneur*, January 29, 2003, https://www.nature.com/articles/bioent710.pdf.

[34] See "Imperial Innovations Group Plc: Year-End Results," Client News, Instinctif Partners, last modified September 13, 2016, http://lifesciences.instinctif.com/news/2016/10/imperial-innovations-group-plc-year-end-results.

[35] See University of Cambridge, *Annual Review 2017* (Cambridge: Cambridge Enterprise, 2017), 5, 23, accessed online, https://www.enterprise.cam.ac.uk/wp-content/uploads/2015/04/updated-Annual-Report-website.pdf.

[36] Lord David Sainsbury, "A Cultural Change in UK Universities," *Science* 296, no. 5575 (June 2002): 1929.

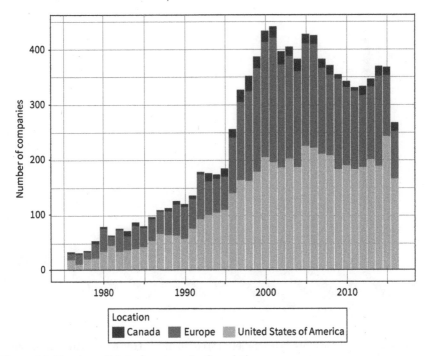

Figure 4.1 Number of biotech companies founded per year
Source: Capital IQ.

Reflecting this desire to match medical innovation with financial returns, the U.K. Medical Research Council (MRC) described its goals as "develop[ing] and sustain[ing] leading edge research programs that will accelerate the transition of fundamental research into measurable positive impact on health, innovation and wealth creation."[37]

The success of those initiatives among academia, various public or private institutions, and industry in turning academic science into entrepreneurial ventures stimulated the interest of a number of countries, which decided to follow suit.[38] Figure 4.1 plots the number of new biotech companies founded each year since 1975 (many but not all of which have academic roots).[39] While

[37] Medical Research Council, *MRC Delivery Plan: 2011/12 to 2014/15*, November 2010, 2, accessed online, https://mrc.ukri.org/publications/browse/delivery-plan-201112-201415/.

[38] For a thorough review of the literature on the performance of university technology transfer offices and additional examples from U.S. and European universities, see Donald S. Siegel, Reinhilde Veugelers, and Mike Wright, "Technology Transfer Offices and Commercialization of University Intellectual Property: Performance and Policy Implications," *Oxford Review of Economic Policy* 23, no. 4 (Winter 2007): 640–60.

[39] The data for the figures of this chapter were obtained from Standard and Poor's Capital IQ database. Our data set included all public and private companies in the Pharmaceuticals, Biotechnology, and Life Sciences industry classifications that are involved in the drug discovery and development process. For

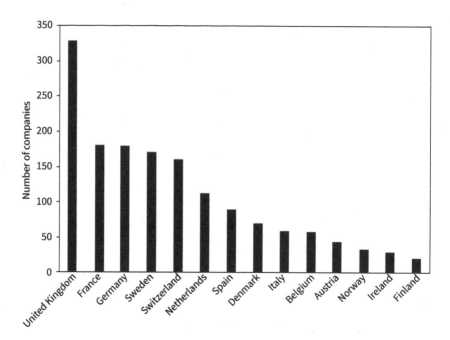

Figure 4.2 Number of biotech (therapeutics) companies in the most active European countries
Source: KPMG.

the early dominance of the United States as the primary home of biotech industry is clearly illustrated, we see that from the mid-1990s European companies become significantly more visible. According to the 2016 Ernst & Young biotechnology report, while the United States continues to be home to more of the world's biotech companies than any other country (with over 2,700 companies in 2015), in Europe alone there were more than 2,250 biotech companies (see Figure 4.2).[40]

The Start-Up Phase of Biotech Companies

These extensive commitments to university technology transfer thus created a hospitable environment for a wide range of potentially valuable discoveries

the purposes of this analysis, we excluded companies that are solely involved in manufacturing (contract manufacturing organizations) or support services (legal, consulting, etc.).

[40] See Ernst and Young, *Beyond Borders: Returning to Earth—Biotechnology Report 2016* (London: Ernst & Young, 2016), 18, 20.

to be developed through the efforts of for-profit companies, but more than science is needed to start and sustain a biotech company. Science's essential partner is a source of funding for the new enterprise, which is typically a venture capital firm.[41] Venture capitalists are professional money managers; that is, they raise a pool of money (a "fund") from investors—often pension funds, university endowments, and other large institutions—and their job is to invest that fund so that its value increases over time. Rather than primarily buying and selling stocks or bonds on the Nasdaq or another stock exchange, they focus on investments in privately held companies—that is, ones generally owned only by the founders, the venture capitalists, and perhaps a few employees.

In some cases, they start the company by providing its "seed capital"; in others, they make "follow-on" investments in existing companies. Unlike the many investors who trade stocks that are listed on the public markets, and who may rarely, or even never, meet someone from a company in which they have invested, venture capitalists generally become intimately involved in a company.[42] Beyond providing the money and the financial expertise in structuring the deal,[43] they help promote the company's "story" to other venture capitalists who could invest in the company in subsequent financing rounds,[44] actively monitor the companies,[45] and in general play a key role in the professionalization of the company, including recruiting a senior management team[46] and introducing the company's scientists to potential corporate partners at large pharmaceutical companies.[47]

[41] The instrumental role of venture capital firms for creating innovative companies has been shown in numerous academic studies: for example, Samuel Kortum and Josh Lerner, "Assessing the Contribution of Venture Capital to Innovation," RAND Journal of Economics 31, no. 4 (Winter 2000): 674–92; and Sampsa Samila and Olav Sorenson, "Venture Capital, Entrepreneurship, and Economic Growth," Review of Economics and Statistics 93, no. 1 (February 2011): 338–49.

[42] For a thorough discussion on the activities that venture capitalists perform, from generating deal flow to selecting opportunities and managing their portfolio companies, see Paul A. Gompers et al., "How Do Venture Capitalists Make Decisions?," Journal of Financial Economics 135, no. 1 (January 2020): 169–90.

[43] See, for example, William A. Sahlman, "Aspects of Financial Contracting in Venture Capital," Journal of Applied Corporate Finance 1, no. 2 (Summer 1988): 23–36; William A. Sahlman, "The Structure and Governance of Venture-Capital Organizations," Journal of Financial Economics 27, no. 2 (October 1990): 473–521; and Steven N. Kaplan and Per Strömberg, "Financial Contracting Theory Meets the Real World: An Empirical Analysis of Venture Capital Contracts," Review of Economic Studies 70, no. 2 (April 2003): 281–315.

[44] See Michael Gorman and William A. Sahlman, "What Do Venture Capitalists Do?," Journal of Business Venturing 4, no. 4 (July 1989): 231–48.

[45] See Josh Lerner, "Venture Capitalists and the Oversight of Private Firms," Journal of Finance 50, no. 1 (March 1995): 301–18; and Steven N. Kaplan and Per Strömberg, "Venture Capitals as Principals: Contracting, Screening, and Monitoring," American Economic Review 91, no. 2 (May 2001): 426–30.

[46] See Thomas Hellmann and Manju Puri, "Venture Capital and the Professionalization of Start-Up Firms: Empirical Evidence," Journal of Finance 57, no. 1 (February 2002): 169–97.

[47] An extensive literature has shown that affiliations with venture capitalists (VCs) are instrumental in attracting corporate partners. See, for example, Toby E. Stuart, Ha Hoang, and Ralph C. Hybels, "Interorganizational Endorsements and the Performance of Entrepreneurial Ventures," Administrative

Great numbers of life sciences discoveries, even patented ones, fail to attract any investment interest, but from time to time, a venture capitalist who hears about an especially interesting technology or potential product may consider starting a new biotech company to take advantage of that discovery. Occasionally venture capitalists first learn of the company from a PowerPoint presentation, business plan, or one-page "teaser" sent by a company, technology transfer office, or founder.[48] Far more likely, someone associated with the company has found a way to network to a member of the venture capital firm, a process by which a personal introduction is likely to increase the chances of at least getting an initial hearing.[49] Companies fortunate enough to get that meeting with a venture capitalist will need to make the most of it, and we discuss the importance of being able to tell a compelling story in our description of the biotechnology entrepreneur in Chapter 6.

To understand the likelihood of venture capital investments, we need to focus briefly on how venture capital firms operate. Much as university scientists compete for a limited amount of funding available from government grants, venture capitalists in the life sciences compete with many other investment firms to raise the funds that they invest in entrepreneurial companies. Venture capital firms are almost invariably partnerships in which a handful of individuals serve as general partners responsible for the day-to-day business of investing in (and sometimes starting) entrepreneurial companies. They invest funds that they have obtained from a group of limited partners, which are typically large institutional investors. Most of these investors are pension funds and university endowments, which, following modern theories of financial portfolio management, place their money in a diverse array of investment opportunities. The bulk of their money is often invested in the

Science Quarterly 44, no. 2 (June 1999): 315–49; David H. Hsu, "Venture Capitalists and Cooperative Start-Up Commercialization Strategy," *Management Science* 52, no. 2 (February 2006): 204–19; Jeffrey J. Reuer and Roberto Ragozzino, "The Choice between Joint Ventures and Acquisitions: Insights from Signaling Theory," *Organization Science* 23, no. 4 (July/August 2012): 1175–90; Umit Ozmel, Jeffrey J. Reuer, and Ranjay Gulati, "Signals across Multiple Networks: How Venture Capital and Alliance Networks Affect Interorganizational Collaboration," *Academy of Management Journal* 56, no. 3 (June 2013): 852–66; and Jeffrey J. Reuer and Roberto Ragozzino, "Signals and International Alliance Formation: The Roles of Affiliations and International Activities," *Journal of International Business Studies* 45, no. 3 (April 2014): 321–37.

[48] Earlier in the history of the industry, every company with a chance to "make a pitch" to a venture capital firm would hand over a long, detailed business plan describing the company's technologies, products, and future plans, along with detailed financial projections covering many years in the future. More recently, those documents have been replaced by PowerPoint presentations conveying much of the same information in bullet point form. See Chapter 5 for a further discussion of these presentations.

[49] In an extensive survey of 885 institutional VCs at 681 firms, Gompers et al. find that the vast majority of new deals comes from the VC's existing networks. Only 10% come inbound from company management. As the authors conclude, "Few VC investments, therefore, come from entrepreneurs who beat a path to the VC's door without any connection." See Gompers et al., "Decisions," 175.

traditional types of stocks and bonds that are listed on various exchanges around the world, but they may elect to put a small portion of their funds in venture capital on the promise of a better financial return than would be expected from investments in less risky and more liquid (that is, more easily bought and sold) financial instruments.

Venture capital firms give an appearance of providing patient, long-term capital to highly promising enterprises. To be sure, their decade-long, often multi-hundred-million-dollar funds provide some degree of financial staying power. Nevertheless, the remarkably high costs of drug development, combined with the ten- to twenty-year timelines associated with the creation of new medicines, mean that venture capitalists rarely provide funding for more than one or two stages—a few years at most—of the company's life. Moreover, the low rates of success in medical research mean that a venture capital fund will spread its investments over a sizable number of companies, perhaps twelve to fifteen companies in any one venture fund. To share the risk and to access more funding, they often invest alongside a small group of like-minded venture capitalists in a "syndicate" assembled for the purpose of participating in what often turns out to be multiple rounds of financing for a company; these rounds are typically called a Series A (B, C, and so on) round. If everything goes well, these lettered rounds generally represent advancing stages of a company's development.

Irrespective of the amount of investment, or whether the investment comes in a start-up Series A round or a substantially later Series D round, the venture capitalists do not seek to achieve their financial returns by receiving dividends paid by the company to its shareholders as a result of the profitable sales of new medicines many years in the future. The day venture capitalists first consider investing in a company is typically the day they begin planning their "exit strategy"—that is, how they will make a good financial return on their investment by selling their shares to someone else. Exits generally manifest themselves in only two forms: a successful IPO that creates a public market for the company's stock, thus allowing the venture capitalists to sell their shares in the stock market (often not until the expiration of a "lock-up" period), or an outright sale of the entire company, usually to a large pharmaceutical company or one of the very large biotech companies, such as Amgen (a "trade sale"). Accordingly, no matter how much the venture capital investors believe in the company's future profitability, they are focused on having the biotech company achieve whatever technical and corporate milestones will create opportunities for a successful exit. In the industry's forty-plus-year history, there have been few cases where venture capitalists have been content to wait for a company to navigate the lengthy process required to successfully develop a new drug.[50]

At the same time, even an earlier exit does not necessarily lead to a positive investment return, and sizable numbers of venture investments in biotech companies are "written off"—that is, all of the money is lost. An analysis of over 1,000 venture capital investments in the life sciences between 2000 and 2010 showed that nearly 60% lost money.[51] It is also worth noting that the investments that turn a profit—because of an exit opportunity for the venture capitalists—will not necessarily develop a successful new medicine. In fact, between 1998 and 2016, only 174 biotech companies originated new medicines that were approved by the FDA.[52]

During the first decade of the twenty-first century, venture capital investments in life sciences have generally ranged from $2 billion to $5 billion per year, and in recent years, there seems to be a significant and consistent increase in the amount of private capital[53] flowing into the biotech industry (see Figure 4.3). In general, while the availability of venture capital is subject to ebbs and flows, it is usually much more stable than the very volatile and unpredictable IPO market. Nevertheless, there are fads and trends in venture investing, as there are in many other aspects of the public and private financial markets.[54] If IPO investors or pharmaceutical companies are buying into fairly early-stage, technology-based companies, so will venture capitalists. Yet often both the public investors and the pharmaceutical companies are looking for somewhat more mature companies—that is, those further along the pathway toward FDA approval. When that is the case, venture capitalists will shift their focus to private companies with more advanced products. At those times, it can be even more difficult to get a new biotech company up and running than in other periods, especially a company with a potential product

---

[50] Those cases would typically occur only when the company itself has been established to further develop a late-stage product that a large pharmaceutical or biotech company has elected to spin off, probably as a result of a shift in the research portfolio strategy.

[51] See Bruce L. Booth and Bijan Salehizadeh, "In Defense of Life Sciences Venture Investing," *Nature Biotechnology* 29, no. 7 (July 2011): 580. The high risk of venture capital investments is also discussed in Robert E. Hall and Susan E. Woodward, "The Burden of the Nondiversifiable Risk of Entrepreneurship," *American Economic Review* 100, no. 3 (June 2010): 1163–94, who find that more than 50% of the venture capital-backed startups had zero-value exits; and William A. Sahlman, "Risk and Reward in Venture Capital," Harvard Business School Background Note 811-036 (December 2010), https://www.hbs.edu/faculty/Pages/item.aspx?num=39710, who finds that 85% of the investors' returns came from just 10% of the investments.

[52] See Chapter 7.

[53] We measured this capital by aggregating all the private placements that took place in biotech companies. The vast majority of this capital comes from VCs, but it also includes investments from corporate venture funds.

[54] For a more general discussion of the impact of public markets on venture capital investments, see Paul Gompers et al., "Venture Capital Investment Cycles: The Impact of Public Markets," *Journal of Financial Economics* 87, no. 1 (January 2008): 1–23.

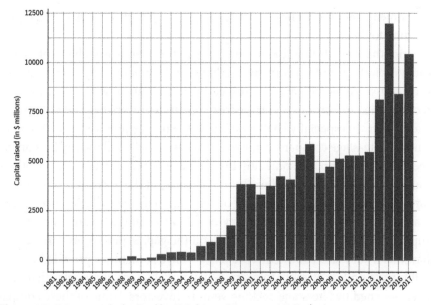

Figure 4.3 Private capital raised by U.S. biotech companies (in $m)
Source: Capital IQ.

or technology still focused on early-stage research. A study shows that such investment cycles have implications not only for the number of companies that receive funding but also for their innovativeness: Companies funded in "hot" equity markets were shown to create more financial value (and even receive higher numbers of patent citations) than companies funded in "cold" markets.[55]

One of us spoke to an early-stage venture capital firm not long after the 2008 crash in the financial markets, and that firm indicated that it had invested in just three of the 1,000 opportunities presented to the firm over a three-year period, which is a depressingly low number for investment-seeking entrepreneurs. Even in relatively good financial times, however, venture capitalists only invest in a small fraction of the opportunities presented to them. Then, of those companies fortunate enough to obtain early-stage, Series A funding, only a fraction of those are able to raise additional funds in a Series B round, and so on. Even though there are billions of dollars of venture capital invested in biotech companies every year, the companies seeking

[55] Ramana Nanda and Matthew Rhodes-Kropf, "Investment Cycles and Startup Innovation," *Journal of Financial Economics* 110, no. 2 (November 2013): 404.

investments far exceed the capacity of the venture capital industry to provide funding to them.

Since there are dramatic ebbs and flows in the overall availability of investment capital for biotech companies, there can be a boom-and-bust feeling in the early-stage biotech arena, irrespective of the rate at which exciting new technologies and products emerge from research universities and other medical centers. In especially challenging times, when venture capital dries up or shifts to late-stage companies, universities have tried to fill the gap by starting seed funds.[56] Although some universities have the resources to do so, others must simply hope that another "boom" cycle is on the horizon. Additionally, "angel" groups—often composed of entrepreneurs who have successfully "cashed out" of companies they previously started and who are interested in supporting new enterprises—also help fill the gap. Super angels—those who have been *very* successful entrepreneurs—have sometimes provided tens of millions of dollars or more to promising young biotech companies that have been spurned by late-stage-seeking venture investors. Some of these companies later raise money from venture capitalists, while others may be acquired, go directly to the public markets in an IPO, or, as with many venture-backed companies, simply go out of business when their products or technology do not sufficiently fulfill their promise by the time the money runs out.

Going Public

A few years ago, one of us was part of a group of biotech CEOs who were asked what their local biotechnology trade organization should do to help them

---

[56] The leaders of Imperial College's technology transfer branch, Imperial Innovations, for example, described their efforts in 2003 as working in "teams [with] a mix of invaluable skills, including expertise in intellectual property assessment, conduct of due diligence, market research, and business development and licensing." Moreover, they noted, "to complement [their] in-house skills, consultants may be engaged, using Imperial College's own seed fund, to carry out a preliminary market study and prepare a draft business plan for the exploitation of [Imperial's] technology"; see Searle, Graves, and Towler, "Biotechnology." A variation on the seed fund theme can be found in a 2003 initiative by Stanford and the University of California to provide funding to advance medical discoveries to the clinical trial stage. Sensing that the pharmaceutical industry and venture capitalists had, at that time, "become more averse to gambling on raw technology and more interested in drugs that have demonstrated at least some promise in a small clinical trial," the universities teamed with SRI International, a research institute, to fill the gap between NIH grant-funded research and the level of development required to attract licensors or investors. See Andrew Pollack, "Three Universities Join Researcher to Develop Drugs," *New York Times*, July 31, 2003, C1. There is no question that in tight financial markets for the biotech industry—and 2001–3 was an especially difficult period—biotech companies and life sciences investors are less aggressive in seeking out new technologies. According to a report in *Nature*, research collaborations between biotech companies and the University of California at San Francisco (UCSF) "peaked at 125 in 2000 for non-clinical research, but then slid to 70 in 2001," and that slide continued through 2002. See Jonathan Knight, "Biotech Woes Set to Hit Academics," *Nature* 418, no. 6893 (July 2002): 5.

build successful companies. Would they like briefings from FDA officials, instruction by experts in organizational behavior and strategic management, scientific networking conferences, and the like? No one objected to any of those suggestions in principle, but the group of seasoned CEOs was unanimous. There was only one thing that they really needed, and that was cash. With enough cash, they could afford all the briefings, instruction, and networking they wanted, but without it, the companies themselves would cease to exist. And since this conversation took place during one of biotech's regular financing slumps in the public markets, corporate life-and-death was an ever-present worry for these executives, as if trying to cure cancer and build a successful enterprise in what may be the world's most highly regulated industry were not enough to give them sleepless nights.

Ultimately, these conversations gave rise to a new state law implementing a creative approach that allows biotech companies to use tax losses to fund research and development—a program that has successfully resuscitated more than one biotech firm that was otherwise on financial life support.[57] But the tax laws will not provide enough funding for the voracious spending needs of biotech companies. Those funds typically have come not only from the venture capital industry, but most importantly, at least in terms of overall cash made available to the industry, also from the public investors—both large institutions and individual investors—that buy stocks and bonds issued by biotech firms listed on the Nasdaq, the London Stock Exchange, Euronext, or other exchanges that have proved hospitable to companies with essentially no revenues and little hope of profits in the foreseeable future. The largest biotech companies are almost invariably publicly held, and while most of the industry, at least as measured by numbers of companies, is composed of privately held companies, most of the money is raised by the minority of companies that become publicly traded. According to an Ernst & Young biotech industry report,[58] in 2013 there were 2,010 private and 339 public companies in the United States. Yet the latter raised almost four times more capital than the former ($19.7 billion compared with $5.6 billion). Accordingly, most of the

---

[57] The result was the New Jersey Technology Business Tax Certificate Transfer Program, which was adopted by the state in 1995. It provided for small biotechnology companies to sell their accumulated tax losses (which they typically were unable to put to use because they had virtually no revenues) to other New Jersey companies, which could employ them to reduce state taxes. The proceeds would allow them to continue developing their products. A study of the program concluded that it "both creates jobs and net tax revenues for New Jersey for biotechnology firms." See New Jersey's Science & Technology University, *Program Evaluation: New Jersey Technology Business Tax Certificate Transfer Program* (Trenton, NJ: New Jersey Economic Development Authority, 2010), 30, accessed online, https://www.nj.gov/trans parency/reports/pdf/NJ%20Technology%20Business%20Tax%20Certificate%20Transfer%20Prog ram%5B1%5D.pdf.

[58] See Ernst & Young, *Beyond Borders: Unlocking Value—Biotechnology Industry Report 2014* (London: Ernst & Young, 2014), 40.

companies that have been successful in the capital-intensive business of creating and developing a new FDA-approved medicine have gone through the IPO process and have accessed (usually repeatedly) large amounts of funding from public investors.

(In this discussion of the public markets, we do not mean to diminish the importance of privately held companies. They have also originated many innovative products, and when they do, they have typically been acquired by pharmaceutical or large biotech companies in the midst of the development process. In fact, their acquisitions via trade sales have been one of the main factors that have sustained the interests of venture capitalists in the life sciences. But the most visible companies, as well as the ones launching new products, have almost invariably been those that have, at some point in their history, become publicly traded.)

Raising money for development-stage biotech companies in the public markets is remarkably unpredictable. The availability of new cash for a publicly held company at virtually any point in its development prior to the commercialization of a new medicine is usually far more dependent upon the mood of investors and the fickle flows of capital in and out of the industry than on the company's fundamental progress toward launching a new product.[59] For people hailing from large pharmaceutical companies—where each year a new allocation of billions of dollars is made available for pouring into research and development—or for those who believe that the financial markets are rational and efficient, this curious financing environment can initially create considerable cognitive dissonance.

A sensible, Business School 101 approach to biotech financing would assume that the highly sophisticated investors dedicated to biotechnology would chart a course from wherever a company is on the drug development pathway, assign a reasonable value to that location based upon some method of evaluating the risks and rewards to be incurred going forward, and then provide increasing amounts of capital at higher and higher valuations as progress is shown (or lower and lower if milestones are missed). And in fact, this is exactly how both companies and investors often talk about potential investments. Although it is more closely approximated in the funding of private companies by venture capitalists, it is a relatively rare occurrence when biotech life in the public markets actually imitates this financial art.

---

[59] See, for example, Michelle Lowry, "Why Does IPO Volume Fluctuate So Much?," *Journal of Financial Economics* 67, no. 1 (January 2003): 3–40. In a cross-industry study, Lowry finds that economic expansion during business cycles and the aggregate level of investor sentiment explain a significant amount of the variation in IPO volume (Lowry, "IPO Volume," 5). Yet both of these drivers are hard to predict ex ante.

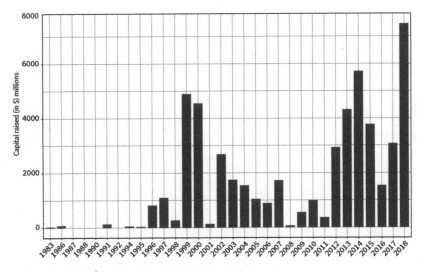

Figure 4.4 Capital raised by U.S. biotech IPOs (in $m)
Source: Capital IQ.

Instead, as Figure 4.4 shows, the availability of capital in the public markets at virtually any stage of development (at least until the company reaches commercial success) is governed by ephemeral financing "windows." The openings and closings of those windows appear to be driven significantly by factors that are not directly related to the industry, and they are impossible to predict. Financing windows open and close with alacrity, little notice, and, in the case of the window slamming shut, frequently much heartache, especially for companies that have devoted months to preparing and launching the IPO process, only to find that prospective investors completely disappeared at the last minute. As one savvy investor said to a biotech CEO hoping to launch an IPO as the market tumbled and the window seemed to be closing, "All I know is that whatever price I would pay for your stock today, it will be worth less tomorrow." Companies forced to abandon an IPO at the eleventh hour not only miss the chance to obtain the cash they need and the liquidity their venture capitalists crave but also get stuck with large legal and accounting bills as well.

During biotech's genome-fueled bull market in 2000, a record (at that time) fifty-eight companies successfully completed IPOs, and, all together, the industry raised over $30 billion, more funding than had been obtained from Wall Street by biotech companies in all of the preceding six years. The next year, after the biotech market suffered a severe downturn, only four U.S. companies were able to "go public," and the industry as a whole raised an amount less than half of 2000's total. A similar pattern can be seen in looking

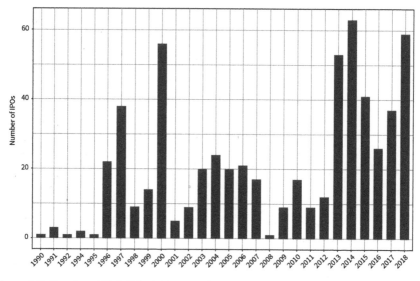

Figure 4.5 Number of U.S. IPOs (1990–2018)
Source: Capital IQ.

retrospectively at biotech windows. The most important historical lesson of the biotech IPO market is that, at least for development-stage companies, it is mercurial at best. Windows of opportunity for IPOs appear without notice, last for an undeterminable amount of time, and then close for indefinite periods (see Figure 4.5).

A variety of explanations have been advanced to explain the volatility of the biotech market, and the opening and closing of financing windows. Rather than advance any particular theory, we will make the pragmatic suggestion that if there were a good explanation that could be consistently applied, then people would be able to predict these events. Yet they remain unpredictable. As a result, managers of public—or hope-to-be-public—biotech companies have no choice but to make business decisions without the ability to predict whether funding will be available at reasonable prices—or at all—when they need it. Companies therefore often try to raise as much money as possible when they can, since they have no reliable ability to predict when they will next be able to do so. This dynamic is nicely summarized in the opening quotation found in Michael Darby and Lynne Zucker's article,[60] where a hypothetical finance professor asks the "new biotech millionaire," "What were the factors

---

[60] Michael R. Darby and Lynne G. Zucker, "Going Public When You Can in Biotechnology," NBER Working Paper Series 8954 (Cambridge, MA: National Bureau of Economic Research, 2002), 1, accessed online, https://www.nber.org/papers/w8954.pdf.

which made it optimal to raise money through an initial public offering at this time?" to receive the not-so-academically-intriguing response: "Because I could."

A biotech company's process for completing an IPO, or "going public," is quite similar to the process followed in other industries. Just like companies in other industries, biotech companies have financial statements audited by a reputable accounting firm, a distinguished board of directors, a credible management team, an attractive business plan, and one or more investment bankers to introduce the company and its "story" to institutional investors and other investment professionals. Together, the company and its bankers—with the assistance of legions of lawyers and accountants—craft a prospectus meeting the strict disclosure requirements of the relevant regulatory bodies such as the Securities and Exchange Commission (SEC) and the Nasdaq in the United States, or perhaps the London Stock Exchange or other similar bodies.

The prospectus is distributed by the bankers to many investment funds. Some of those funds specialize in biotechnology and other healthcare investments, while others are "generalists" who have, at least in hot biotech markets, devoted a portion of their investment portfolios to biotech stocks. The funds are then offered the opportunity to meet with the management team in large and small meetings during a tortuous two- or three-week process, during which the company's senior executives meet with prospective investors from dawn to dusk each day. These frequently involve group presentations during breakfast and lunch with one-on-one meetings scheduled all morning and afternoon, followed by a flight to the next city where the process will be repeated the next day. This "road show" continues until pricing, the day when the IPO shares are allocated to the investors who have ordered them. Trading of the newly public shares begins soon thereafter, with anxious executives and bankers hoping for the stock to trade well—that is, go up—in the post-IPO trading known as the "aftermarket." A few days later, the closing takes place, and the company deposits the IPO proceeds in its bank account, less a healthy underwriting discount to the bankers (that is, the bankers' fees generally take the form of a discounted price the underwriters pay for the shares, which they immediately resell to investors at the IPO price) and a slew of legal, accounting, and other costs.

The two initial principal parties in the IPO process are the company and the investment bankers. Companies seeking to accomplish an IPO need to find a willing bank, and bankers, who make a considerable amount of money via brokering IPOs and other financings, are anxious to find suitable IPO candidates. This should be the easy part. But it is not. For obvious reasons, many biotech companies want to execute an IPO as soon as possible whenever the window is open. In hot markets, this number may represent a sizable portion of the private companies at the time. It is highly unlikely that all of them will have the opportunity for at least two reasons: (1) the market has sometimes lost interest after a spate of biotech IPOs, and (2) there are a limited number of IPOs that can be effectively handled by the fairly small number of investment banks with the staff, expertise, and track record to consummate IPOs in the biotech investment sector. As a result, the leading banks can accommodate the desires of only a select number of the companies that would like to go public in any given year. Also, given the volatile nature of the biotech IPO market, it is unreasonable to expect that any financing window will stay open indefinitely. Thus, many companies compete for the attention of relatively few investment banks; meanwhile, the banks are competing vigorously to win the IPO mandate from those companies that appear to be the hottest prospects.

What makes a company a hot prospect? It is whatever the bankers believe will be the easiest deal to sell to investors, and that can change from window to window. When the genome was being mapped, any company with a "genomics story" became a hot prospect. About a year later, genomics had cooled off to such an extent that genomic IPOs were essentially untenable. In some IPO markets, products in late-stage clinical testing seem to be required, while at other times good laboratory data tied to an exciting new technology have been adequate. And at any time, investments from well-regarded venture capitalists, combined with attractive collaborations with major pharmaceutical companies, can provide comfort to investment bankers and investors that knowledgeable third parties have analyzed the company in detail and found the technology or product exciting enough to make an alliance worthwhile.

For a hot prospect, the issue is not whether a banker will be interested but which banker to select. Even before the company contemplates an IPO, bankers will visit the company and its venture capitalists, provide them with research reports and other market intelligence, and so on. When the opportunity for an IPO arises, these hot-prospect companies typically hold a "beauty contest," in which teams of bankers from several leading banks are invited to make a brief presentation to the key executives and perhaps the board of directors.

The beauty contest process requires a series of nearly identical bankers from virtually indistinguishable banks to distinguish themselves from their competitors crowding the company's small reception area. Among the firms that are most active in completing biotech financings, they all are likely to have more than adequate sales, trading, research, and banking capacity to complete an IPO successfully when the window is open. And since personnel turnover in all of these arenas is rampant, it can sometimes seem to managers and board members as if the banks all employ the same people. The major point of comparison occurs in the "league tables" presented by each firm in its "pitchbook," invariably created by the most junior banker, who stayed up all night putting his or her (usually his) Ivy League or Oxbridge education to use preparing the materials to be used by more senior bankers and who thereafter remains mute while trying desperately not to fall asleep while the senior bankers make their pitch.

The league tables are designed to show that this particular bank is the Number One Biotech Investment Bank. And, indeed, in a five-bank beauty contest, each bank will rank first. This trick is accomplished by making sure that the ranking measures whatever the bank argues should be the most relevant accomplishment, defined as whatever elevates that particular bank to a number one position, and is arguably defensible as a measurement of banking success. Thus, the bank that participated in the most IPOs will rank by the number of deals, while another bank will choose to rank by the total amount of money raised if that bank did not participate in as many deals but took part in the largest ones. A third bank might choose "aftermarket performance" (that is, how well its IPO stocks traded in the market after the deals were launched). One smaller, niche banker presented itself to one of us as being number one at being the second most important bank in biotech financings involving multiple bankers. In other words, they pitched themselves as being the #1 Number Two Bank.

The IPO provides the company with a chance to access new sources of capital while allowing founders and early investors (typically, venture capitalists) to "get liquid"—that is, to have the opportunity to convert shares of stock into cash, sometimes a very large amount of cash. One trade journal has also kept track of newly minted "molecular millionaires," biotech executives whose stock and stock options have made them at least "paper" millionaires, especially during robust stock market cycles that can make biotech stock certificates very valuable.[61]

Although most biotech IPOs have taken place on America's Nasdaq stock market, biotech companies can now be found listed on the stock exchanges in many countries. In fact, many countries' bourses have either relaxed their listing requirements or have formed smaller, more flexible markets in hopes of attracting capital for high-tech investment within their borders. It is, therefore, possible (subject, of course, to market conditions) for a company to

---

[61] See, for example, Alex Philippidis, "Top 10 Molecular Millionaires," *GEN: Genetic Engineering & Biotechnology News*, January 2, 2017, http://www.genengnews.com/insight-and-intelligenceand153/top-10-molecular-millionaires/77899901.

launch an IPO in numerous countries without any past or present revenues or profits (and in many cases, without any reasonable expectation of revenues or profits in the foreseeable future). Most public biotech companies have traditionally been listed in the United States, Canada, and Western Europe, but it is increasingly common to find biotech companies listed on an exchange in almost every industrialized country. Companies do not necessarily list on an exchange in their home countries, nor are companies limited to one listing. A number of the larger European companies, for example, have chosen to list both in their local market and on Nasdaq, which remains the global leader in providing the most active and liquid market for the shares of biotech companies.

Conclusion

Biotech company formation requires a combination of money and molecules, joined together in the pursuit of new medicines. At any one time, there are over 5,000 biotech companies around the world, ranging from virtual companies, one-person entities focused on a single early-stage product, to complex, hierarchical organizations with thousands of employees and robust product sales. In this chapter, we have tried to outline the key elements of starting and raising money for these companies, and in the next chapter we will turn to their symbiotic relationships with the larger pharmaceutical companies through the formation of alliances.

Before doing so, it is worth noting that while traditionally there seemed to be a rather clear distinction between the role played by early-stage investors (such as venture capitalists) and late-stage development partners (such as the pharmaceutical companies), the dividing line between them seems to be increasingly blurry. The vast majority of pharmaceutical companies have now established corporate venture capital (CVC) units that make substantial investments in early-stage companies. The existing academic research on CVCs has generally explored cross-industry data sets that have a heavy emphasis on information technology and therefore do not account for the unique characteristics of the drug development process.[62] Accordingly, our

---

[62] See, for example, Paul Gompers and Josh Lerner, "The Determinants of Corporate Venture Capital Success: Organizational Structure, Incentives, and Complementarities," in *Concentrated Corporate Ownership*, ed. Randall K. Morck (Chicago: University of Chicago Press, 2000), 17–54; Markku Maula and Gordon Murray, "Corporate Venture Capital and the Creation of US Public Companies: The Impact of Sources of Venture Capital on the Performance of Portfolio Companies," in *Creating Value: Winners in the New Business Environment*, ed. Michael A. Hitt et al. (Oxford: Blackwell Publishers, 2002), 164–87; and Gary Dushnitsky and Michael J. Lenox, "When Do Incumbents Learn from Entrepreneurial

understanding of how those CVC units operate, how they perform relative to independent venture capital funds, and the role they play (if any) in their pharmaceutical sponsors' research and development efforts remains rather limited. With the growing interest of pharmaceutical companies in early-stage companies, seen in the establishment of these CVC units, combined with our recommendations in Chapter 7 for the pharmaceutical industry to look for ways to access more innovation from the biotech industry, we believe that this a promising area for future research.

Nevertheless, there is an open question as to the extent to which biotech companies will need (or want) increased interactions with pharmaceutical companies at the earliest stages of research and development, especially in light of the growth of contract research organizations (CROs) that are capable of performing virtually all of the tasks involved in drug development. CROs have allowed a substantial number of biotech companies to be fully capable of proceeding to late-stage development and even to the market without the need for partnerships with large pharmaceutical companies.[63] Those recent trends cast doubt on the traditional assumptions regarding the symbiosis between the biotech industry's focus on research and the pharmaceutical companies' expertise in drug development, and they highlight the need for continuing research on the constantly evolving biopharmaceutical universe.

Ventures? Corporate Venture Capital and Investing Firm Innovation Rates," *Research Policy* 34, no. 5 (June 2005): 615–39.

[63] See, for example, Ben Adams, "Biotechs Getting Bigger in Late-Stage R&D, Leaving Big Pharmas Behind: Report," *Fierce Biotech*, April 23, 2019, https://www.fiercebiotech.com/biotech/biotechs-getting-bigger-late-stage-r-d-leaving-big-pharmas-behind-report.

5

Biotech–Pharma Alliances

In 2008, Myriad Genetics licensed to the Danish pharmaceutical company H. Lundbeck the European rights to Flurizan, a product that had successfully completed Phase II trials, and a Phase III trial was well underway. Only the final stages of that trial would need to be completed before the companies could file for regulatory approval. Flurizan, a member of the ibuprofen family of anti-inflammatory molecules, was being developed by the American biotech company to treat Alzheimer's disease, and Lundbeck had long been an expert in that field. Originally founded early in the twentieth century, Lundbeck was an international pharmaceutical company with a well-known specialization in the development and commercialization of products for central nervous system (CNS) diseases, such as Alzheimer's. It had introduced its first CNS product, for schizophrenia, over half a century before.

To acquire the European rights to Flurizan, Lundbeck entered into a "corporate partnership" with Myriad that called for a series of payments, including $100 million upon the signing of the agreement, with $250 million more due in connection with various regulatory approvals. Lundbeck also agreed to pay cash bonuses upon the achievement of various commercial milestones, together with very substantial royalties (20% to 39% of all sales of Flurizan). Not only would Myriad receive these payments from Lundbeck's activities in Europe, but the biotech company would also retain the commercial rights to Flurizan in the rest of the world, including the large and lucrative U.S. market.[1]

Every year, there are typically a thousand or more deals like this, with biotech and pharmaceutical companies seeking synergistic ways of working together. This volume of biotech–pharmaceutical agreements, both in the aggregate and on an annual basis, is impressively high, and even in the early days of those alliances scholars have suggested that the sheer numbers of these arrangements may be "without precedent in business history."[2] Since

---

[1] See Myriad Genetics, "Myriad Genetics Selects Lundbeck as European Partner for Flurizan," news release, May 22, 2008, https://investor.myriad.com/static-files/e02cacd4-b32a-486a-9e46-338126339823.

[2] Frank T. Rothaermel, "Complementary Assets, Strategic Alliances, and the Incumbent's Advantage: An Empirical Study of Industry and Firm Effects in the Biopharmaceutical Industry," *Research Policy* 30, no. 8 (October 2001): 1240.

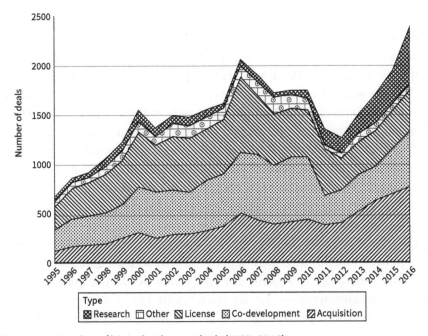

Figure 5.1 Number of biotech–pharma deals (1995–2016)
Source: Recombinant Capital (now part of Clarivate Analytics).

then, the number of those alliances in a given year has more than tripled (see Figure 5.1).[3]

Biotech companies offer innovative products and technologies that can contribute to filling the pharmaceutical companies' pipelines,[4] and pharma companies can provide biotech companies with much needed cash[5] and, in many cases, drug development expertise[6] as well. In light of the fickle quality

[3] The data for this chapter come from the Recombinant Capital (RECAP) alliance database (now part of Clarivate Analytics). Our analysis includes all deals that took place in the biopharma industry during the period 1995–2016, except for deals that were exclusively focused on distribution, manufacturing, marketing, or generics. As a "co-development" deal, we categorized any deal that included the keywords co-development/collaboration/joint venture in its description. As a "Research" deal, we categorized any deal that included the keyword "research" but didn't make any reference to licensing, co-development, or acquisition. Lastly, in our calculation of the average deal sizes, we excluded "mega-deals" (for example, deals with total value above $3 billion as such outliers would skew our estimations).

[4] See Gary P. Pisano, "The R&D Boundaries of the Firm: An Empirical Analysis," *Administrative Science Quarterly* 35, no. 1 (March 1990): 153–76; Rothaermel, "Complementary Assets," 1235–51; and Matthew J. Higgins and Daniel Rodriguez, "The Outsourcing of R&D through Acquisitions in the Pharmaceutical Industry," *Journal of Financial Economics* 80, no. 2 (May 2006): 351–83.

[5] See Josh Lerner and Robert P. Merges, "The Control of Technology Alliances: An Empirical Analysis of the Biotechnology Industry," *Journal of Industrial Economics* 46, no. 2 (March 2003): 125–56; and Josh Lerner, Hilary Shane, and Alexander Tsai, "Do Equity Financing Cycles Matter? Evidence from Biotechnology Alliances," *Journal of Financial Economics* 67, no. 3 (March 2003): 411–46.

[6] See Ashish Arora and Alfonso Gambardella, "Complementarity and External Linkages: The Strategies of the Large Firms in Biotechnology," *Journal of Industrial Economics* 38, no. 4 (June 1990): 361–79; Frank T. Rothaermel and David L. Deeds, "Exploration and Exploitation Alliances in Biotechnology: A System

of the equity financing markets for biotech,[7] the lively market for these kinds of biotech–pharma alliances has proved to be an essential way for biotech companies to try to satisfy their nearly insatiable need to raise enough cash to keep their products moving toward FDA approval and commercial sale.

Over the years, these agreements between biotechnology companies and pharmaceutical companies have provided many billions of dollars to the biotechnology companies. At the same time, they have transferred the commercial rights to many potential new drugs and drug discovery technologies from the biotech industry to pharmaceutical companies or, in some cases, from small biotech companies to larger ones, such as Celgene or Amgen, that have already successfully commercialized new products and are seeking to add to their own product pipelines.

The basic concept underlying these transactions is that the biotech company has developed a technology or potential product that the pharmaceutical company[8] would like to acquire, essentially on a pay-as-you-go installment plan. These agreements—which are often referred to by the parties as "corporate partnerships" or "strategic alliances"[9]—come in all shapes and sizes, from fairly minor transactions involving bits of technology peripheral to the biotech company's primary focus to major deals that can transform both parties: the small biotech company may, for the first time, have enough resources to develop the product, and the large pharmaceutical licensee could be adding a commercial blockbuster to its therapeutic pipeline.

These agreements also involve companies across the entire spectrum of different technologies, from traditional small molecules (which we will sometimes call "synthetics") to gene therapy. As can be seen in Figure 5.2, while all technologies demonstrate an increasing trend in terms of sheer number of deals, the ratio of biologics has increased over time. A remarkable increase over the last couple of decades can also be noted in oncology deals, which capture almost half of the entire market in terms of volume in recent years (see Figure 5.3).

of New Product Development," *Strategic Management Journal* 25, no. 3 (March 2004): 201–21; Frank T. Rothaermel and Warren Boeker, "Old Technology Meets New Technology: Complementarities, Similarities, and Alliance Formation," *Strategic Management Journal* 29, no. 1 (January 2008): 47–77; and J. Adetunji Adegbesan and Matthew J. Higgins, "The Intra-alliance Division of Value Created through Collaboration," *Strategic Management Journal* 32, no. 2 (February 2011): 187–211.

[7] See Lerner, Shane, and Tsai, "Equity Financing Cycles," 411–46.
[8] While the licensee may be a larger biotech company, to make the descriptions less complicated, we will refer to all the licensees in these transactions as pharmaceutical companies.
[9] Ranjay Gulati, "Alliances and Networks," *Strategic Management Journal* 19, no. 4 (April 1998): 293–317. For an analysis of biotech strategic alliances from the perspective of social theory, see Mark de Rond, *Strategic Alliances as Social Facts: Business, Biotechnology & Intellectual History* (Cambridge: Cambridge University Press, 2003).

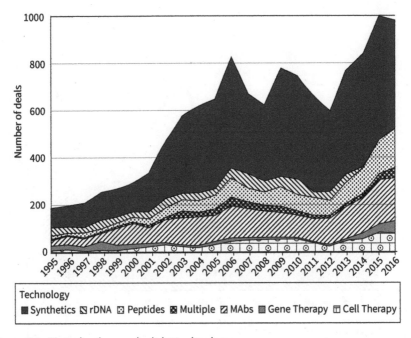

Figure 5.2 Biotech–pharma deals by technology
Source: Recombinant Capital (now part of Clarivate Analytics).

Because the largest pharmaceutical companies tend to have the most money and the biggest R&D pipelines, they are responsible for a sizable number of the alliances every year. In Figures 5.4 and 5.5, we show the breakdown of the portfolio of alliances done by nine major pharmaceutical companies by technology category (for example, recombinant DNA) and by principal therapeutic area from 1996 to 2016. The ratios (denoting the number of deals in a specific technology/area over all deals) are remarkably consistent across all nine companies.

In some years, the aggregate amount of cash raised by biotech companies through these alliance arrangements can exceed the total amounts received by the industry from all other sources,[10] including sales of equity to venture capitalists and other investors. As a result, these types of agreements have been one of the critical components in the growth of the biotechnology industry.[11]

[10] See Sean Nicholson, Patricia M. Danzon, and Jeffrey McCullough, "Biotech-Pharmaceutical Alliances as a Signal of Asset and Firm Quality," *Journal of Business* 78, no. 4 (July 2005): 1433–64; and Stacy Lawrence and Riku Lahteenmaki, "Public Biotech 2013—The Numbers," *Nature Biotechnology* 32, no. 7 (July 2014): 626.
[11] See Walter W. Powell, Kenneth W. Koput, and Laurel Smith-Doerr, "Interorganizational Collaboration and the Locus of Innovation: Networks of Learning in Biotechnology," *Administrative Science Quarterly* 41, no. 1 (March 1996): 116–45; Lerner and Merges, "Technology Alliances," 125–56; Nicholson, Danzon, and McCullough, "Biotech-Pharmaceutical Alliances," 1433–64; Nadine Roijakkers and John Hagedoorn,

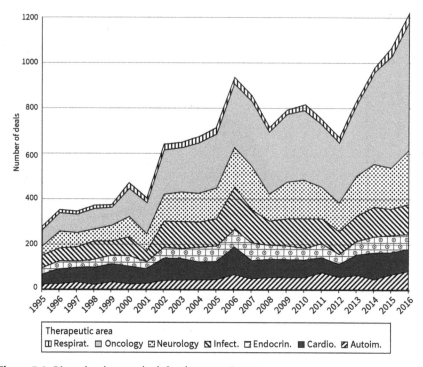

Figure 5.3 Biotech–pharma deals by therapeutic area
Source: Recombinant Capital (now part of Clarivate Analytics).

It may be worth noting that there are many ways to calculate the value of these alliances, and the headlines rarely capture the inevitable contingencies. In press releases and other public announcements, these deals are frequently described in terms of the total amount of all potential fixed payments—that is, upfront payments, equity purchases, guaranteed research funding and milestones, but not royalties or shared profits. Industry insiders commonly call these total amounts "Biobucks" or "BioWorld Dollars" since that is how they tend to appear in the headlines of the industry newsletter *BioWorld Today.* The accounting firm Ernst & Young noted in a 2017 report that the $3.5 billion in upfront cash received in the 2016 alliances had potential values to the biotech companies of nearly 57.7 billion BioWorld Dollars.[12] Over the

"Inter-firm R&D Partnering in Pharmaceutical Biotechnology since 1975: Trends, Patterns, and Networks," *Research Policy* 35, no. 3 (April 2006): 431–46; and Jaideep Anand, Raffaele Oriani, and Roberto S. Vassolo, "Alliance Activity as a Dynamic Capability in the Face of a Discontinuous Technological Change," *Organization Science* 21, no. 6 (November/December 2010): 1213–32.

[12] See Ernst & Young, *Beyond Borders: Staying the Course—Biotechnology Report 2017* (London: Ernst & Young, 2017), 84–85. Similarly, public announcements typically describe almost any biotech–pharma

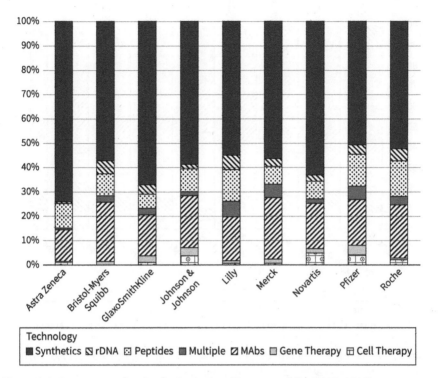

Figure 5.4 Percentage of technologies in the alliance portfolio (1996–2016)
Source: Recombinant Capital (now part of Clarivate Analytics).

years, the upfront payments have tended to be in the range of 6–12% of the total number of BioWorld Dollars involved in the deal,[13] while the average deal size has increased considerably in the past decade for most technologies and particularly so for monoclonal antibodies (see Figure 5.6). It is important to remember, especially when reading the glowing language of the press release announcing the deal, which almost inevitably includes the BioWorld Dollar value of the transaction, that the pharma company can usually terminate the alliance at virtually any time and is thus freed from making any remaining payments (except perhaps for a "wind-down" fee).

contractual arrangement as a partnership, collaboration, or alliance, yet some alliances involve little more than a simple license of patents or technology from one party to another, even if they are described by the parties as alliances or partnerships. Nevertheless, these transactions are typically based on a product or technology that has been created either by the biotech company itself or by the company founder's academic research laboratory. And during the alliance, it would not be unusual for there to be an element of technology transfer from the biotech company to the pharmaceutical company, followed by some degree of shared responsibility for the future development of the product or technology, although the extent to which the companies engage in genuinely collaborative research varies widely.

[13] See Ernst & Young, *Industry Report 2017*, 93–94.

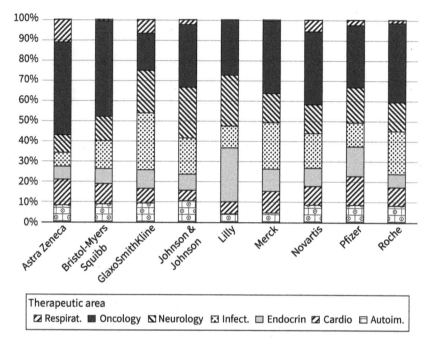

Figure 5.5 Percentage of therapeutic areas (TAs) in the alliance portfolio (1996–2016)
Source: Recombinant Capital (now part of Clarivate Analytics).

The Principal Deal Terms

While each of these partnering transactions has elements tailored specifically to the product or technology involved, they tend to follow a reasonably
common format.[14] The financial terms frequently include:

(1) An initial "upfront" payment by the pharmaceutical company, some-
    times accompanied by an equity investment, in which the pharma
    company purchases some of the biotech company's stock, often at a
    premium to the market price for public companies, or the most recent
    venture capital round for private ones.
(2) Committed research funding that will be used to pay the biotech
    company to keep the product moving forward. In some transactions,
    the pharma company will agree to work on the product in its own

[14] See David T. Robinson and Toby E. Stuart, "Financial Contracting in Biotech Strategic Alliances,"
Journal of Law and Economics 50, no. 3 (August 2007): 559–96.

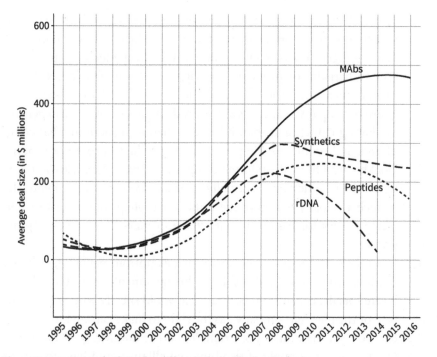

Figure 5.6 Average deal size for different technologies (in $m)
Source: Recombinant Capital (now part of Clarivate Analytics).

laboratories as well, or it may pay for clinical trials to be performed at hospitals around the country or around the world. In some cases, the pharmaceutical company agrees to pay all remaining development costs; in others, the biotech company will pay a share of those costs.

(3) There will almost always be milestone payments to the biotech company upon the achievement of specified research and development goals, such as initiating Phase I, Phase II, or Phase III clinical trials, filing with FDA, and so on. For products that could be used for a variety of different disease indications—multiple types of cancer or autoimmune diseases, for example—there may be a new set of milestones for each disease-related program. There may even be "post-commercialization" milestones that become triggered by the achievement of particular levels of annual revenues, such as the first time the global sales reach $500 million or $1 billion.

(4) The pharmaceutical company will pay royalties to the biotech company, or the two companies will share in the profits (if any) from sales of the product. Royalties represent a fixed percentage of the sales of

the product, and it is not unusual for them to increase as the annual revenues increase, on the theory that the product becomes more profitable at higher levels of sales. Alternatively, the biotech company may seek to bargain for profit-sharing rights in place of royalties. Doing so may mean that the smaller company will need to subsidize the pharma company's expenses as the product goes through an expensive marketing launch, but then it has the chance to share in what could ultimately become sizable profits if the product becomes a genuine blockbuster. Often, the pharma company will have the exclusive rights to sell the product, but, especially in profit-sharing arrangements, the biotech company will have the right—or perhaps the obligation—to create a sales force so as to participate in marketing and selling the product.

(5) In addition to the payment terms such as upfront payment, research funding, milestones, and royalties, the agreement would typically allocate specific control rights. That is, it would specify whether the biotech or the pharma company has the right to make certain kinds of key decisions in the future. For example, Lerner and Merges[15] have analyzed a random sample of two hundred alliances and have identified as many as twenty-five control rights that, depending on the companies' relative bargaining positions, may be allocated either to the pharmaceutical company or the biotech company. These control rights can determine anything from the management of the clinical trials (for example, termination decisions) to intellectual property (for instance, ownership of core or derivative technologies) to the alliance governance structure, including whether the pharmaceutical firm is assigned a tie-breaking vote, equity stake, and corresponding voting rights. The allocation of such control rights is particularly important in environments that entail significant uncertainty because in those environments not all future contingencies can be anticipated and therefore accounted for in the initial contract.[16]

[15] See Lerner and Merges, "Technology Alliances," 125–56.

[16] See Sanford J. Grossman and Oliver D. Hart, "The Costs and Benefits of Ownership: A Theory of Vertical and Lateral Integration," *Journal of Political Economy* 94, no. 4 (August 1986): 691–719; Oliver Hart and John Moore, "Property Rights and the Nature of the Firm," *Journal of Political Economy* 98, no. 6 (December 1990): 1119–58; and Philippe Aghion and Jean Tirole, "The Management of Innovation," *Quarterly Journal of Economics* 109, no. 4 (November 1994): 1185–209.

Valuation Analyses

Not surprisingly, at the outset, there is often an impressively widespread gap between what the biotech company thinks its "crown jewel" is worth and what a pharmaceutical company is willing to pay. Typically, both companies start with two basic analytical tools: a valuation of the product based on a prediction of its ability to generate cash flows in the future and a "comparables" analysis based on the terms of similar transactions that have been entered into by other companies and that involve (arguably) comparable products.

The discounted cash flow (DCF) analysis and its more technical variation called real-option valuation,[17] which accounts for probabilistic scenario analysis, are based on the modeling techniques generally taught in business schools; they can be used to value products (or entire companies) in virtually any area in business. These kinds of models typically start with the optimistic assumption that the product development efforts will move forward through all the stages of research and development—on plan and at the budgeted costs—and that the product will be approved by the FDA and then rapidly launched in the major world markets. Based on assumptions (many of which later turn out to have been wildly inaccurate) about the size of the market, the degree of market penetration, the forecasted price of the product, the likelihood of termination at specific stages of research and development, and so on, the product's future cash flows are predicted over as much as a ten- or fifteen-year period after the product's launch, meaning that the financial model may extend for twenty years or more into the increasingly hazy future. Then, the expenses are added, as the model seeks to account for all of the future cash needed for development, launch, marketing, manufacturing, and other necessary expenses. The remaining net cash flows (that is, inflows minus outflows) are discounted heavily to account for the cost of capital and for the potential risk of failure. In biotech product models, it is not uncommon for a discount rate of 20% or more per year to be used, meaning that a dollar of net cash flow in one or two decades is only worth pennies today. A consulting firm reports that, in 2011, a "survey of 242 biotech professionals with valuation experience . . . found an average discount rate of 40.1% for early-stage projects, 26.7% for mid-stage projects and 19.5% for late-stage projects."[18]

---

[17] See Ralph Villiger and Boris Bogdan, "Getting Real about Valuations in Biotech," *Nature Biotechnology* 23, no. 4 (April 2005): 423–28.

[18] Jonathan Stasior, Brian Machinist, and Michael Esposito, *Valuing Pharmaceutical Assets: When to Use NPV vs rNPV* (London: Alacrita Consulting, 2018), 3, accessed online, https://www.alacrita.com/whitepapers/valuing-pharmaceutical-assets-when-to-use-npv-vs-rnpv. They cite Ralph Villiger and Nicolaj Hoejer Nielsen, *Discount Rates in Drug Development* (Basel, Switzerland: Avance, 2011), accessed online, https://pdfs.semanticscholar.org/c91f/88118fd68982b3211ebe61b9c6232c1d9983.pdf.

Ultimately, this complex analysis leads to a "net present value" (NPV), or the risk-adjusted net present value (rNPV) in the case of a real-option analysis, of the product—basically, the value today of owning all of the future net cash flows. The deal terms, in theory, should be based on this net present value, which, at the outset, belongs entirely to the biotech company. The theory is that the biotech company owns an asset currently worth an amount that can be calculated—the NPV—and if the pharma company wants to acquire that product, a fair price would be a series of payments whose NPV is the same (or more). The payments to be made when the agreement is signed are valued on a dollar-for-dollar basis, while the milestone payments during development, as well as the royalties after commercial launch, are discounted on the same basis as the future expenses.

Calculating the NPV is often made more complicated by the array of patented technologies—not all of which are necessarily owned by the biotech company—that may have been involved in creating the product. The biotech company, therefore, often owes some of those future cash flows to universities or other companies from which it has acquired licenses for technology used in the product. More often than not, there are several of these "upstream" licenses.

One key aspect of the biotech–pharma negotiation revolves around which party will pay the "stacking royalties"[19] associated with these upstream licenses—that is, if the biotech company has agreed to pay a 1% royalty for technology A, plus 2% for technology B, and 3% for C, for a total of 6%, and the pharma company proposes to pay only a 10% royalty, the biotech company will end up with less than half of the total royalty payments. To avoid this problem, biotech companies will try to shift the burden of these stacking royalties to the pharma company as much as possible.

All of these NPV calculations look very scientific on paper, with sales and costs predicted to the penny many years into the future. Despite the apparent mathematical precision of these models, however, the calculations generated by the two parties are likely to be widely disparate. The models are capable of great precision, but in the end, the valuations are driven most powerfully by the assumptions, which is where the parties inevitably diverge, usually dramatically. The biotech company's assumptions will be uniformly optimistic—speedy clinical development and rapid regulatory approval with a broad label indicating that a large part of the market, not just a narrow subset of the sickest

---

[19] Michael A. Reslinski and Bernhard S. Wu, "The Value of Royalty," *Nature Biotechnology* 34, no. 7 (July 2016): 685–90.

patients, could potentially benefit from the product, all of which leads to a future blockbuster.

Meanwhile, the pharmaceutical company will make consistently conservative assumptions: long and expensive clinical trials, marginally positive results, only a narrow label approval from the FDA, and a highly competitive marketplace, with only modest market share. The two companies may even disagree about pricing and cost of goods, all culminating in dueling models—each dozen of pages long and based on exactly the same modeling techniques—that have almost no points of intersection and that lead to NPVs that differ by hundreds of millions of dollars or more.

A great deal of effort goes into these models, and arguing about them may be a nearly ritualistic component of negotiating an alliance, but ultimately, they are likely to be cast aside in favor of one or both of the two factors that are much more likely, from a practical point of view, to drive the business discussions: the "comparables" analysis and the relative bargaining power of the biotech compared with that of the pharma. During a lengthy session discussing a pharma company's DCF analysis, for example, one of us was able to point to an error in the model, one that would increase the value of our product considerably. The junior analyst responsible for the model promised to correct the mistake and to circulate a revised version to both companies. Immediately, the large company's senior-most negotiator jumped in to say, "Sure, he can fix the model, but we're still not going to pay you any more."

Far more relevant in that negotiation, and in many such discussions, is an analysis of what "comparable" deals look like. Although biotech companies typically speak of the uniqueness of their innovative science and their pathbreaking products, an alliance negotiation frequently includes a considerable degree of haggling over what accurately represents the alliance "market" for similar products. With over one thousand publicly announced biotech–pharma deals each year, there is an active alliance marketplace in which it can be possible to align various characteristics of this particular product with comparable products at similar stages of development at the time of the alliance. Comparability is often assessed based on the disease indication (cancer, inflammation, infectious diseases, etc.) and the stage of development. Accordingly, a "comparables analysis" of a cancer product entering Phase II clinical trials could include recent deals involving other products for cancer in Phase I/II trials or those beginning Phase III trials, even if those products were based on an entirely different technology and would be used for different cancers. Alternatively, when certain areas of technology become highly desirable—and there are fads in drug development, as there are in every field—a comparables analysis might include deals relating to other products based on the "hot" technology, such as other monoclonal antibodies or gene therapies, irrespective of disease focus.

Imagine, for example, that Antibody Biotech Company (ABC) has a monoclonal antibody–based product that has just completed a Phase I/II trial in thirty patients with lung cancer and is talking with potential partners about an alliance. In considering a possible deal structure, it is likely to pull together information about cancer antibody alliances over the past few years (see Table 5.1).

As with the DCF analysis, the deals included in the comparables analysis may vary considerably between the biotech and pharma lists. One side has an incentive to find a way to include the richest possible deals and, in a search for the "best" comparables, might reach back several years or include products only tangentially related to the one involved in the hoped-for alliance. The other side, of course, seeks to do the opposite, including as many data points as possible from transactions where the licensee was able to secure a very attractive deal. Nevertheless, the two comparables analyses generated by the companies will usually more closely resemble each other than the DCF calculations and are therefore more useful in moving the parties toward a mutually satisfactory transaction. Additionally, alliances are often major events in the life of a biotech company, and since the principal terms are usually announced in a press release, both parties want to be perceived as getting a reasonable deal in the context of the alliance market prevailing at that time.

Table 5.1 Clinical-stage antibody deals (2007–2008)

Licensor/ Licensee	Clinical stage	Lead indication	Total value (in millions of dollars)	Upfront (in millions of dollars)	Royalties
Seattle Genetics/ Genentech	I/II	Hematologic cancers	860	60	Double digit
Macrogenics/Lilly	II/III	Diabetes	300	41	Double digit
Tolerx/GSK	End of II	Diabetes	760	70	Double digit
Kyowa Hakko/Amgen	I	Cancer	520	100	Double digit
Bioinvent/Roche	End of I	Cancer	775	77.5	10–12%
Immunomedics/ Nycomed	II	Inflammation/ Autoimmune	620	40	10–16%

How the Terms Are Set

While the discounted cash flow calculations and the comparables analysis are some of the standard tools most frequently used in the valuation process, perhaps the most important factor in determining where the terms of a particular deal may fall within (or even outside of) the "market" range is the relative bargaining power of the biotech company compared with that of the pharmaceutical company. Given the need of the biotech company for cash and other resources to keep moving its products toward FDA approval, its bargaining power is usually determined primarily by two factors: the availability of alternative financing options—such as the public and private equity markets[20]—and the extent to which multiple pharmaceutical companies are competing to acquire the product or technology in question.[21]

Beyond the variability in the external funding environments, the ratio of supply and demand for specific technologies and products is bound to be a key determinant of the outcome of the negotiations. In an environment in which pharmaceutical companies routinely need to supplement their internal research efforts with additional products, it would seem likely that a "bidding war" among multiple pharma companies would be a feature of most alliance negotiations. As it turns out, many—probably a substantial majority—of these biotech–pharma partnerships are not the result of a competitive bidding process in which the biotech company essentially selects its preferred pharma partner. To understand why this is the case, it is important to see the alliance marketplace as, in fact, a series of smaller, more discrete, markets that have very different supply and demand ratios.[22]

The most obvious "sellers' market" involves transactions where a biotech company has found the resources to develop a potential blockbuster product in a commercial market of strategic interest to many large pharmaceutical companies to the point where enough clinical safety and efficacy have been shown to point to a high likelihood of FDA approval. For example, Higgins[23] finds that pharmaceutical companies that engage in alliances involving later-stage products tend to transfer more rights to the biotech company than those alliances with lead products in earlier stages of development.

[20] See Lerner, Shane, and Tsai, "Equity Financing Cycles," 411–46; Matthew J. Higgins, "The Allocation of Control Rights in Pharmaceutical Alliances," *Journal of Corporate Finance* 13, no. 1 (March 2007): 58–75; and Toby E. Stuart, Salih Zeki Ozdemir, and Waverly W. Ding, "Vertical Alliance Networks: The Case of University-Biotechnology-Pharmaceutical Alliance Chains," *Research Policy* 36, no. 4 (May 2007): 477–98.

[21] See Pisano, "R&D Boundaries," 153–76; and Adegbesan and Higgins, "Intra-alliance Division," 187–211.

[22] Adegbesan and Higgins, "Intra-alliance Division," 187–211.

[23] Higgins, "Control Rights," 58.

Moreover, subsequent research has shown that in late-stage alliances, and particularly those where a limited number of biotech companies seem to fit the pharmaceutical companies' need for assets, the biotech companies tend to appropriate more value through retaining more control rights.[24] The fit with the pharmaceutical company's pipeline is often a shifting landscape, as pharmaceutical companies periodically decide to get into or out of markets such as cancer or cardiovascular disease.[25] Products such as Flurizan—most of the way through Phase III, and showing good clinical results in a huge potential market—likely had multiple pharma "suitors" and therefore could command a deal involving not only many BioWorld Dollars but also a considerable amount of real ones, in this case, in the form of a $100 million upfront payment for only the European rights.

These deals—where the biotech company has several eager bidders and considerable negotiating power—do not occur as frequently as might be expected because so many biotech companies enter into alliances at earlier stages of development (see Figure 5.7). For much of the industry's history (and until fairly recently), it has been nearly impossible for a small company to secure enough funding to be able to develop a product all the way to Phase III trials. In most cases, the cash contributions of a pharmaceutical partner are essential, and even when the equity markets have embraced the industry with enthusiasm, investors have often resisted funding biotech companies that do not have the positive "quality signal" of a partnership with a large pharmaceutical company.[26] For example, Nicholson et al. find that biotech firms that have formed an alliance receive substantially higher valuations from venture capitalists and other investors at subsequent financing rounds.[27] They argue that this finding is consistent with a signaling model (that is, a sign of quality—sort of a "Good Housekeeping Seal of Approval") as well as the presence of "asymmetric information in financial markets, such that pharmaceutical companies are better able than money managers to evaluate the scientific and managerial expertise of private biotech firms."[28]

---

[24] See Adegbesan and Higgins, "Intra-alliance Division," 204.

[25] See Rita Gunther McGrath and Atul Nerkar, "Real Options Reasoning and a New Look at the R&D Investment Strategies of Pharmaceutical Firms," *Strategic Management Journal* 25, no. 1 (January 2004): 1–21; and Umit Ozmel and Isin Guler, "Small Fish, Big Fish: The Performance Effects of the Relative Standing in Partners' Affiliate Portfolios," *Strategic Management Journal* 36, no. 13 (December 2015): 2039–57.

[26] See Powell, Koput, and Smith-Doerr, "Interorganizational Collaboration," 116–45; Toby E. Stuart, Ha Hoang, and Ralph C. Hybels, "Interorganizational Endorsements and the Performance of Entrepreneurial Ventures," *Administrative Science Quarterly* 44, no. 2 (June 1999): 315–49; and Nicholson, Danzon, and McCullough, "Biotech-Pharmaceutical Alliances," 1433–64.

[27] See Nicholson, Danzon, and McCullough, "Biotech-Pharmaceutical Alliances," 1433.

[28] Nicholson, Danzon, and McCullough, "Biotech-Pharmaceutical Alliances," 1437.

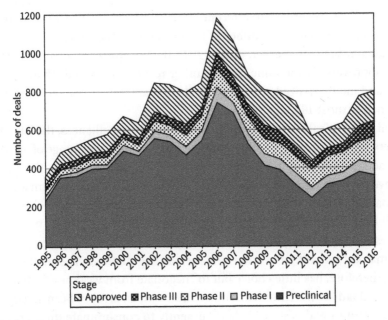

Figure 5.7 Deals by stage of development
Source: Recombinant Capital (now part of Clarivate Analytics).

Table 5.2 The antibody-partnering landscape as of 2008

Phase of development	Already partnered/acquired	Unpartnered
Preclinical	15	40
Phase I (or I/II) ongoing	14	14
Phase II ongoing	17	5
Phase III ongoing	9	1

To see how often small biotech companies were able to "hang onto" their products to the stage of Phase III clinical trials, we analyzed the monoclonal antibody–based products being developed by biotech companies as of 2008 that were in clinical trials or advanced preclinical development (that is, products that would soon be in clinical trials). Of the total of 115 products, there was only one unpartnered product that had reached Phase III clinical trials. In fact, there were not many unencumbered Phase II products, as shown in Table 5.2.

Interestingly, recent evidence seems to suggest that while unpartnered products were a rarity up to a decade ago, this may not necessarily be the case anymore. According to a recent industry report,[29] since 2014 there has been a steady rise in the number of biotech companies launching their own products in the U.S. market, and in 2018 those companies filed 27% of the FDA approvals. At the same time, another recent report finds that, in 2018, large and medium-sized pharmaceutical companies were responsible for less than half of those FDA approvals.[30]

Still, in a typical year featuring a thousand or more alliances, the late-stage "sellers' market" deals typically represent only a small percentage of all of the transactions, and they are hardly representative. In most cases, the parties' relative negotiating position is likely to be reversed, with the pharmaceutical companies picking and choosing among thousands of biotech opportunities.[31] Here, too, the biotech company will try to solicit the interest of multiple parties, but for a variety of reasons, there is often only one interested potential pharma partner willing to enter into serious negotiations. The smaller company, therefore, has little choice but to "negotiate from weakness"—that is, it will try to radiate the confidence that comes with an array of licensing opportunities, but in reality it needs fairly urgently to consummate this particular transaction. At the same time, the pharmaceutical company can rarely be sure that it is "the only game in town," which makes it reluctant to try to extract extremely one-sided terms, especially since large pharma companies generally try to maintain a positive image as a "partner of choice" so that they are well positioned to attract the best opportunities where there is, in fact, a bidding war.

While there often seems to be an inexhaustible supply of biotech-generated product opportunities, the demand side of the equation features a much smaller number of "buyers," and most of the buying power is concentrated in the top handful of pharma companies, which are responsible for spending the lion's share of the industry's R&D expenses each year. These are the companies that dominate the alliance market, and each will enter into scores of transactions each year. When those companies suffer setbacks in their internal pipelines, they will more aggressively seek to in-license new opportunities and will be inclined to agree on terms more favorable for the biotech company

[29] See Erin McCallister, "Independents' Day: After Their First Solo Launch, Biotechs Look to New Partnering Strategies," *BioCentury*, May 3, 2019, https://www.biocentury.com/article/302071/how-the-next-wave-of-independent-companies-are-adapting-for-long-term-growth.

[30] See IQVIA Institute for Human Data Science, *The Changing Landscape of Research and Development: Innovation, Drivers of Change, and Evolution of Clinical Trial Productivity* (Parsippany, NJ: IQVIA, 2019), 5, accessed online, https://www.iqvia.com/insights/the-iqvia-institute/reports/the-changing-landscape-of-research-and-development.

[31] See Richard Mason and Donald L. Drakeman, "Comment on 'Fishing for Sharks: Partner Selection in Biopharmaceutical R&D Alliances' by Diestre and Rajagopalan," *Strategic Management Journal* 35, no. 10 (October 2014): 1564–65.

to ensure they get what they want. In fact, the largest companies tend to provide the most generous terms in their alliances, as shown in Figure 5.8. By the same token, when two large pharmaceutical companies merge—as they have had a tendency to do—that can cause both companies to decrease their activities in the "partnering market" while the surviving company sorts out the newly combined portfolio of alliances and internal research programs. When these kinds of internal factors affect one or more of the largest pharmaceutical companies, the effects can be felt throughout the alliance marketplace.

Most of the preceding discussion has been focused on the financial terms, as they are typically the primary motivation for the biotech company to transfer all or part of the rights of one of its few product opportunities

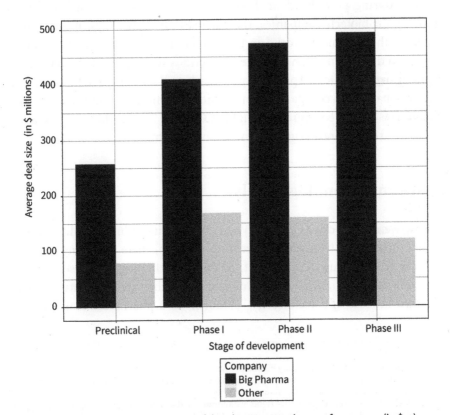

Figure 5.8 Average deal size by stage of development and type of company (in $m)

Note: Big pharma companies included in the calculation of this figure are Abbott, AbbVie, AstraZeneca, Bayer, Bristol-Myers Squibb, GlaxoSmithKline, Johnson & Johnson, Lilly, Merck, Novartis, Pfizer, Roche, Sanofi, and Takeda. Future research into these alliances may be able to determine the extent to which the large companies paying these premiums receive greater value in terms of the numbers of regulatory approvals or the amount of commercial sales generated by successful products.

Source: Recombinant Capital (now part of Clarivate Analytics).

to another company. But a considerable amount of time, effort, and bargaining—and much of the lengthy final agreement—is devoted to the issue of control rights. The theoretical models of incomplete contracts and control rights[32] predict that the value created by the alliance would be greater if the biotech firm retains more control rights, especially for the earlier stages of research, because its knowledge about the product or the technology is likely to have a greater marginal impact on the success than that of the pharmaceutical company's. Yet a study of biotech–pharma alliances has shown that, in practice, the pharmaceutical company normally retains the vast majority of control rights, especially when the biotech is in a weaker funding position.

Does this mismatch between theory and practice matter? The answer appears in a subsequent study,[33] which shows that, with respect to alliances formed during periods when there is a limited availability of external financing for biotech companies, the alliances not only assign more control rights to the pharmaceutical firm but also are less likely to lead to an FDA-approved drug than alliances signed in periods with a greater availability of funding in the public and private equity markets.[34] It is not clear, however, that the pharmaceutical companies' commitment to centralized internal decision-making (as discussed at greater length in Chapter 7) will allow them to agree to assign critical control rights to small biotech companies.

Platform Technology Deals

Lastly, we should clarify that, beyond the pipeline-filling, product-related deals, such as the transaction involving Flurizan, there are many alliances formed around drug development "enabling" technologies. These deals are typically called "alliances" or "partnerships," but they are, in reality, straightforward licenses of a technology "platform" to a pharmaceutical company

---

[32] See Aghion and Tirole, "Management of Innovation," 1185–1209; and Lerner and Merges, "Technology Alliances," 125–56.

[33] See Lerner, Shane, and Tsai, "Equity Financing Cycles," 411–46.

[34] Interestingly, in a related stream of work, Higgins finds that financial markets respond more favorably (measured by the cumulative abnormal returns accruing to the pharmaceutical firm) to the announcement of a strategic alliance when fewer control rights are allocated to the pharmaceutical company. See Higgins, "Control Rights," 74. Cumulative abnormal returns (CARs) are a standard measure of determining the effect of specific events (for example, lawsuits, buyouts, or an announcement of CEO succession) on stock prices. They are estimated by summing (over a small window of time, often three to five days) the differences between the daily expected return of the stock (what it should have been without the event) from the actual return. The cumulative abnormal returns to the pharmaceutical company may result from investors' appreciation of the enhanced likelihood of building future value through the alliance because more control rights were retained by the biotech company, or perhaps because the effervescent market for biotech investing has been accompanied by increased volatility in the shares of pharmaceutical companies that announce deals with biotech companies.

that wants to use that technology in its own research efforts. We call these "cash-and-carry" deals because the pharmaceutical company often takes the technology away and uses it in its R&D labs to create a new product. Take, for example, monoclonal antibody technology, the key elements of which (for a variety of historical reasons) belonged to a handful of biotech companies. Those companies—primarily Abgenix, Cambridge Antibody Technology, Medarex, and Protein Design Labs—had a two-pronged strategy for employing their antibody-creating technologies: (1) they would generate their own products by figuring out which disease targets would be ideal to manipulate using antibodies, and (2) they would license their technology to other companies that had different ideas about attractive targets. Each of these monoclonal antibody technology companies entered into numerous (in some cases dozens of) alliances, some of which were cash-and-carry licenses while others were product-related alliances surrounding products that were initially developed by the antibody companies and then "partnered" with pharmaceutical companies, usually during the course of clinical development.

Cash-and-carry deals occupy a different portion of the alliance market than product-related deals. They are, virtually by definition, very early-stage deals, as the technology represents the initial platform from which a product candidate is fashioned. Because there are so many of these cash-and-carry deals involving a wide range of technologies, their presence may skew analyses of the broader product alliance market, and biotech companies seeking a partnership around an early-stage antibody *product* will avoid including cash-and-carry deals in the comparables analysis since the value proposition is so different. While cash-and-carry deals often have milestones and royalties, both the real and the BioWorld dollars are usually much lower than for product-related transactions. Product companies will argue that cash-and-carry deals are like buying paintbrushes, whereas they have painted the next Mona Lisa. The cash-and-carry response, of course, is that, without the brush, the product company just has a blank sheet of paper. To understand where a particular transaction fits within the biotech–pharma alliance market, it is important to appreciate whether the deal involves a cash-and-carry "tool" (like the paintbrush) or the new therapeutic product itself, on its way to blockbuster/masterpiece status.

Does One Party Have an Edge?

So what happened to Flurizan? Right at the outset, the biotech company received $100 million in much-needed cash. Lundbeck, however, came up empty-handed. Just six *weeks* after the deal was signed, Flurizan failed to achieve satisfactory Phase III results and was abandoned by both companies.[35]

With experiences like this, it is not surprising that, in light of all of the academic studies focusing on biotech–pharma alliances, some economists have invoked the famous "lemons problem" to describe biotech–pharma alliances. As Nobel laureate George Akerlof first described in 1970 in a discussion about the automobile market, the only conceivable time that used car sellers will offer a bargain price is when the sellers know something bad about the cars that they hope buyers will not discover. Economists call this phenomenon, where the parties' ability to know the value of the product is likely to be unequal, "information asymmetry."[36] That is, one party to the transaction has a much better ability to assess the true value of the item being sold (or out-licensed) than the other.

A number of scholars studying biotech–pharmaceutical alliances have argued that the biotech licensors have more information about the value of their products (or any underlying technology on which they are based) than the pharmaceutical licensee.[37] For these observers, it seems obvious that the technology-based company will be in a much better position to understand the value of its technological assets than anyone else. And if biotech companies really do enjoy such an information benefit, there is likely to be a lemons problem; that is, biotech companies will offer only lousy products and technologies to the pharmaceutical industry, ones that, like Flurizan, will end up as big disappointments.

This assumption essentially places biotech–pharma licensing transactions on one rung of a virtual ladder of information asymmetries in the drug development industry. The deepest knowledge, at least for new technologies,

[35] See Andrew Pollack, "Myriad Genetics Stops Work on Alzheimer's Drug," *New York Times*, July 1, 2008, https://www.nytimes.com/2008/07/01/business/01gene.html.

[36] Nancy T. Gallini and Brian D. Wright, "Technology Transfer under Asymmetric Information," *RAND Journal of Economics* 21, no. 1 (Spring 1990): 147–60; and Stefan Wuyts and Shantanu Dutta, "Licensing Exchange—Insights from the Biopharmaceutical Industry," *International Journal of Research in Marketing* 25, no. 4 (December 2008): 273–81.

[37] See Gary P. Pisano, "R&D Performance, Collaborative Arrangements and the Market for Know-How: A Test of the 'Lemons' Hypothesis in Biotechnology," working paper, Harvard Business School, Cambridge, MA, July 1997; Lerner and Merges, "Technology Alliances," 125–56; David B. Audretsch and Maryann P. Feldman, "Small-Firm Strategic Research Partnerships: The Case of Biotechnology," *Technology Analysis & Strategic Management* 15, no. 2 (June 2003): 273–88; Manel Antelo, "Licensing a Non-Drastic Innovation under Double Information Asymmetry," *Research Policy* 32, no. 3 (March 2003): 367–90; Rothaermel and Deeds, "Exploration and Exploitation," 201–21; Nicholson, Danzon, and McCullough, "Biotech-Pharmaceutical Alliances," 1433–64; Gary P. Pisano, *Science Business: The Promise, the Reality, and the Future of Biotech* (Brighton, MA: Harvard Business School Press, 2006); Richard Mason, Nicos Savva, and Stefan Scholtes, "The Economics of Licensing Contracts," *Nature Biotechnology* 26, no. 8 (August 2008): 855–57; and Umit Ozmel, Jeffrey J. Reuer, and Ranjay Gulati, "Signals across Multiple Networks: How Venture Capital and Alliance Networks Affect Interorganizational Collaboration," *Academy of Management Journal* 56, no. 3 (June 2013): 852–66.

originally resides within the universities whose faculty members have invented or discovered the technology and whose licensing offices have spawned much of the industry.[38] The biotech company, which ultimately becomes the licensor in the alliances with pharmaceutical companies, is often formed in technology transfer transactions in which the venture capital investors may be in a weaker position to assess the value of the underlying discoveries than the inventors. These venture-backed companies then further develop the technology, leading to a strategic alliance with a pharmaceutical company, which, in turn, may have less information than the company in whose laboratories the technology resides and which may be less able to understand the technology than the biotechnology company. And then the alliance itself provides a positive signal to even less well-informed equity investors, who may purchase the biotech company's stock at least in partial reliance on the validation of the company's technology, products, and drug development capabilities found in the endorsement of a relatively more knowledgeable pharmaceutical company. If this scenario is true, then every new group of investors—often contributing increasingly large amounts of funding—knows less about the company and its products than everyone who has come before.

Why would anyone participate in such a lack-of-information pyramid? An important factor is that different types of investors may have different goals. Venture capitalists and other equity investors acquire a biotech company's stock in the hope that someone else will buy it from them at a higher price than they paid, typically long before the final clinical results are in. Venture capitalists plan to sell their stock to the public following an IPO or to a pharma company in an acquisition. IPO investors hope new public market investors will get excited enough about the company to bid the price of the stock up, something that frequently happens when a big pharma alliance is announced, and so on. For many investors, the positive signal created by having a large pharmaceutical company interested in the company may be far more valuable in and of itself—that is, as a sign that big pharma is interested and therefore might pay a great deal to buy the entire company in the future—than as a predictor of the ultimate commercial success of the product many years later.

For example, one study shows that privately held biotech firms with prominent alliance partners are able to go public sooner and earn greater valuations at the IPO than firms without such ties. This study specifically highlights the role of signaling by empirically demonstrating that "much of the benefit of having prominent affiliates stems from the transfer of status

---

[38] See, for example, Stuart, Ozdemir, and Ding, "Vertical Alliance Networks," 477–98, on the brokerage role played by biotech companies between universities and pharmaceutical companies.

that is an inherent byproduct of interorganizational associations."[39] A related stream of literature finds that the announcement of a strategic alliance by a public biotech company leads to higher shareholder value measured by the higher cumulative abnormal returns that occur the days following the announcement.[40]

Unlike the equity investors, the pharmaceutical company that is in-licensing a product is the one party in this chain of arguable asymmetry that really does invest in the product's success in treating human disease. It would thus be most disadvantaged by information asymmetry and therefore least likely to be an active participant in the alliance market, at least in theory. Yet the opposite is the case, and we believe that it is because drug development is such an uncertain proposition that deep knowledge of the technology does not necessarily provide better insights into the ultimate value of the product. Put differently, nobody knows in advance which new treatments will be safe and effective new medicines, no matter how much they know about the molecules. Well-known statistics show that approximately 90% of all pharmaceutical products entering human clinical trials ultimately fail to achieve regulatory approval.[41] The likelihood of failure is even greater for the average product involved in a licensing transaction because many such transactions take place at the even riskier preclinical stage.[42]

The reality is that both companies suffer from a symmetrical lack of information, not necessarily about the technology, but about whether any particular application of it will safely and effectively intervene in a disease pathway in a way that will lead to a commercially important new medicine. Consider the following counterexample to the Flurizan case, which can help to illustrate the point that the biotech licensor is not necessarily in a better position than the licensee to predict the product's ultimate value.

About a decade before the Myriad–Lundbeck Flurizan deal, biotech company Idec Pharmaceuticals entered into a licensing transaction with Genentech, which, at that time, was a twenty-year-old commercial-stage biotech company. (It is now a wholly owned subsidiary of the international pharmaceutical company Roche.) The primary product involved in the

---

[39] Stuart, Hoang, and Hybels, "Interorganizational Endorsements," 315.

[40] See Su Han Chan et al., "Do Strategic Alliances Create Value?," *Journal of Financial Economics* 46, no. 2 (November 1997): 199–221; and Higgins, "Control Rights," 58–75.

[41] See Steven M. Paul et al., "How to Improve R&D Productivity: The Pharmaceutical Industry's Grand Challenge," *Nature Reviews Drug Discovery* 9, no. 3 (March 2010): 203–14; Michael Hay et al., "Clinical Development Success Rates for Investigational Drugs," *Nature Biotechnology* 32, no. 1 (January 2014): 40; and Joseph A. DiMasi, Henry G. Grabowski, and Ronald W. Hansen, "Innovation in the Pharmaceutical Industry: New Estimates of R&D Costs," *Journal of Health Economics* 47 (May 2016): 20–33.

[42] See Nicholson, Danzon, and McCullough, "Biotech-Pharmaceutical Alliances," 1452.

transaction, named C2B8, was a monoclonal antibody for the treatment of a malignancy of white blood cells known as non-Hodgkin's lymphoma. The product had completed Phase II clinical trials (that is, it had been evaluated for both safety and initial signs of efficacy), and it was about to enter the final phase of clinical testing prior to filing for FDA approval. It was not quite as far down the development pathway as Flurizan had been at the time of the Lundbeck deal, but it was fairly close, in that both Idec and Genentech had the opportunity prior to entering into the alliance to review the data from clinical testing of the product in lymphoma patients.

Under the terms of the alliance, Genentech acquired a greater than 50% commercial interest in the product, plus an option to acquire interests in two additional products being developed by Idec, all for initial payments totaling $9 million. Five million dollars of that upfront payment would take the form of a purchase of Idec's preferred stock by Genentech, and the larger company committed to purchase an additional $17.5 million of Idec stock prior to FDA approval of C2B8. Finally, Genentech agreed to pay Idec up to $30.5 million in milestone and option payments. In short, for a fraction of the amounts Lundbeck would later pay for the European rights to Flurizan (that is 57 million BioWorld Dollars for C2B8 versus 350 million BioWorld Dollars for Flurizan), Genentech acquired over half the global commercial rights to multiple product opportunities, as opposed to just the European rights to Flurizan acquired by Lundbeck. It is worth noting, however, that at the time of the Idec–Genentech alliance, monoclonal antibody–based products had not yet become highly successful commercial products for cancers and lymphomas; therefore, the likelihood of success was probably not seen as being very high. Moreover, the modest potential market for lymphoma patients may have appeared much smaller than the commercial opportunities offered by a treatment for the much larger numbers of patients with Alzheimer's disease.

In any event, the C2B8 product, now called Rituxan, was approved by the FDA two years later and has become one of the biopharmaceutical industry's most successful blockbuster drugs, thanks not only to its use in treating lymphoma but also its use in a number of autoimmune diseases. Genentech's total investment of less than $60 million (a considerable portion of which was in the form of equity purchases, which meant that Genentech received potentially valuable Idec shares on top of its commercial interest in C2B8) rapidly returned hundreds of millions of dollars, and eventually billions of dollars, of annual profits from the sales of C2B8/Rituxan. Worldwide sales of Rituxan from 2012 to 2017 were approximately $7.5 billion per year, making it not only one of Roche's bestselling products ever but also the second-highest

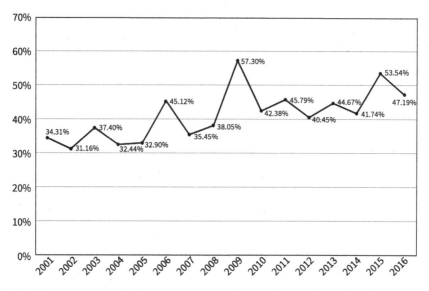

Figure 5.9 Percentage of in-licensed drug candidates in the pipeline of big pharma
Source: Thomson Reuters Cortellis.

selling medicine in the world in 2017.[43] Using any reasonable assumptions for calculating net present value, the value of the rights acquired by Genentech/Roche far exceeded the value of the payments obtained by Idec. It is hard to imagine that Idec would have sold the rights at that price if it had the benefit of special (otherwise hidden) insights into the product's future value.

Meanwhile, from the licensees' perspective, it is clear that the pharmaceutical companies have conquered their fears of buying medicinal lemons. There is a robust market of one thousand alliances a year. In fact, as we can see in Figure 5.9, the percentage of in-licensed drugs in the collective R&D pipeline of the top fifteen largest pharmaceutical companies (by sales in 2016) demonstrates an increasing trend from 2001 onward.

As we argue in this chapter and subsequent ones, the answer to the big question—will this product be safe and effective in treating disease?—is basically unknowable based on the current state of scientific knowledge. Neither the pharmaceutical company nor the biotechnology company is in any sort of privileged position when it comes to assessing the ultimate likelihood of success or failure. As a result, it would seem likely that the success rates of in-house and in-licensed products would be the same.

---

[43] See Alex Philippidis, "The Top 15 Best-Selling Drugs of 2017," *GEN: Genetic Engineering & Biotechnology News*, March 12, 2018, https://www.genengnews.com/a-lists/the-top-15-best-selling-drugs-of-2017/.

Some earlier research argued that in-house-developed drugs are more likely to succeed than in-licensed, but subsequent research showed that in-licensed drugs might have a higher likelihood of success.[44] Yet as the authors of the latter study acknowledge, because their data set does not specify the exact timing for the in-licensed products, their superior performance might be driven by timing issues alone (for example, in-licensed deals taking place in a late stage of the development process). A study in which two of us participated, along with our colleague Arthur Neuberger, overcomes this methodological challenge. It explicitly focuses on the pipelines of biotech companies. The study controls for the timing issue and concludes that there is no statistically significant difference between the success rates for in-licensed drugs and those of in-house ones.[45]

Alliance Terminations

Given the 90% failure rate of clinical-stage products, it is not surprising that these transactions are typically terminated by the large company in fewer than four years.[46] Moreover, at the preclinical stages, when many transactions take place, the likelihood that the product will ultimately succeed in obtaining FDA approval may be as low as 0.01%.[47] At this point, we should pause to note that alliances are not always terminated because the product has failed some sort of critical test and is therefore no longer worth developing further toward commercialization. Sometimes that is the case, as it was with Flurizan, but it is not unusual for the pharmaceutical company to withdraw from an alliance for any of a number of reasons, some of which have little or nothing to do with the product.[48] The pharmaceutical company may have changed its overall R&D strategy; other products that have been internally developed or in-licensed may appear to address larger markets; or perhaps the company needs to cut its budget, and the alliance is too expensive. Irrespective of what the press releases call these transactions—alliances, licenses, collaborations,

[44] See Patricia M. Danzon, Sean Nicholson, and Nuno Sousa Pereira, "Productivity in Pharmaceutical–Biotechnology R&D: The Role of Experience and Alliances," *Journal of Health Economics* 24, no. 2 (March 2005): 317–39; Ashish Arora et al., "A Breath of Fresh Air? Firm Type, Scale, Scope, and Selection Effects in Drug Development," *Management Science* 55, no. 10 (October 2009): 1638–53; and J. A. DiMasi et al., "Trends in Risks Associated with New Drug Development: Success Rates for Investigational Drugs," *Clinical Pharmacology & Therapeutics* 87, no. 3 (March 2010): 276.
[45] See Arthur Neuberger, Nektarios Oraiopoulos, and Donald L. Drakeman, "Lemons, or Squeezed for Resources? Information Symmetry and Asymmetric Resources in Biotechnology," *Frontiers in Pharmacology* 8, no. 338 (June 2017): 1–4.
[46] See Pisano, *Science Business*, 155.
[47] See Rothaermel and Deeds, "Exploration and Exploitation," 209.
[48] See McGrath and Nerkar, "Options Reasoning," 1–21; and Ozmel and Guler, "Small Fish," 2039–57.

partnerships, etc.—the vast majority of them are really simply option agreements. That is, the pharmaceutical company can ultimately obtain the commercial rights if the product is successfully approved in return for a series of predetermined payments. Meanwhile, at any time, the pharmaceutical company can cancel the agreement, and there are usually lengthy provisions that deal with what happens to the product in the event of such a termination. Often the product is simply returned to the licensor, sometimes accompanied by a termination fee, and the original owner can then continue the product's development if it chooses to do so.

Information Symmetry

At the outset of an alliance, both parties have high hopes of success, despite the fact that only a tiny percentage of products ultimately survive the daunting drug development odds. Neither party knows whether this product will be one of the few success stories. The impressively high likelihood of failure associated with products involved in biotech–pharmaceutical alliances thus creates an environment in which there is, paradoxically, an unusually high degree of symmetry between the parties as to the ultimate value of the product. In applying the lemons hypothesis to biotech alliances, researchers have made an unstated assumption that superior knowledge about the goods (that is, the nature of the technology) correlates positively with a better understanding of the *value* of those goods (for example, the likelihood that any molecule will become a commercially successful medicine). This assumed link—or even equivalence—between technical knowledge and the predictability of commercial value does not necessarily apply to biotech–pharmaceutical licensing transactions. The licensor and licensee may have different opinions about the likelihood of success, the time and cost of further development, the strength of patent protection, the size of the market, and other common drug development variables, but it is not clear that there is knowledge available to the licensor that puts it in the position of the quintessential example of information asymmetry, the used car seller—that is, someone who possesses otherwise hidden technical knowledge that provides the seller with a distinct advantage over the buyer in assessing the value of the item being sold.[49]

---

[49] An exception here would be a case in which the biotech company used fraudulent data to license an otherwise inert or fictitious product. The extensive due diligence efforts typically pursued by the pharmaceutical companies should minimize or eliminate this possibility. See Russell W. Coff, "How Buyers Cope with Uncertainty When Acquiring Firms in Knowledge-Intensive Industries: Caveat Emptor," *Organization Science* 10, no. 2 (March/April 1999): 144–61.

In short, in something as unpredictable as drug development, there is rarely a chance to have a secret advantage. There are so many unknowns in the development process that neither buyer nor seller is likely to have an asymmetrical edge. This inability of all parties to predict the future commercial value of a potential drug accurately is not limited to early-stage programs. It can be seen even in the licensing of products that have completed nearly all stages of development, including a considerable amount of human testing, as can be seen in both the Flurizan flop and Rituxan's resounding success.

In this environment, biotech–pharma licensing transactions might best be understood as taking place under conditions of nearly complete information symmetry, not because it is possible for both parties fully to share and to comprehend all information necessary for ascertaining the true value of the product, but, to the contrary, because *neither* party is capable of obtaining such information. At the time of the typical biotech–pharmaceutical alliance, the true economic value of the product or technology being bargained for is equally unknowable by either party. They may differ as to their perceptions of the value, but neither is likely to have information that puts it in the position of having significantly better insight into that value than the other.

This symmetrical lack of information about the value of the licensed product can be seen in how the deals are structured. While pharmaceutical companies will sometimes acquire biotech companies to gain complete ownership of their development-stage products, licensing transactions typically contain a stream of contingent payments. Additionally, even some acquisitions include earn-outs, which means that the selling shareholders get only part of their money at the time of the sale. The rest of the purchase price is paid only when the key product (or products) achieves important milestones, such as FDA approval. These kinds of structures can serve as a tool to allow a less knowledgeable licensee to deal with problems of adverse selection caused by information asymmetry,[50] but in biotech–pharmaceutical licensing, the same structure can be seen as serving both parties' strategic interests. A licensee may prefer "back-loaded" royalties and other contingent payments to mitigate against the risk of loss associated with a product's early failure (see the Myriad–Lundbeck example discussed earlier). At the same time, a licensor could find the same structure attractive because it is worried about "selling too cheap" if it gives up a future participation in billions of dollars of profits in a blockbuster drug (see the Idec–Genentech example discussed earlier).

---

[50] See Coff, "How Buyers Cope," 144–61; and Mason, Savva, and Scholtes, "Licensing Contracts," 855–57.

We should note here that this skepticism about whether any of the parties possess better information does not necessarily mean that nothing relevant to the R&D process is knowable. Quite to the contrary, the results of experiments, the quality of the existing data, the pace of development, and many other crucial elements of the drug development process are both knowable and can be evaluated equally well by pharma and biotech companies. Not surprisingly, the detailed scientific due diligence analysis conducted by pharma scientists (sometimes with the help of leading academic scientists) can extend over months and can even include in-house scientific work. There is a great deal to be analyzed, reviewed, and evaluated. These analyses are important not only to help the pharma companies make their decisions but also to create quality signals to the financial markets, where investors often do not have the resources—or access to confidential data—to perform the same type of comprehensive analysis.

Conclusion

Because of the very high cost of drug development, relatively few biotech companies are able to access sufficient amounts of equity capital to maintain full ownership of their products until successful commercialization. Even among the publicly traded companies, resources are limited. The vast majority are unprofitable; around 25% of them operate with less than a year's worth of cash, while another 25% have less than two years' worth of cash.[51] In 2008, for example, two-thirds of biotech companies in the United States had less than two years of cash, largely because the equity financing markets were diminished for much of the time following the technology bubble of 2000. Unlike the owner of a perfectly running used car, who has no reason to sell, especially into a market that discounts its value for defects that might be hidden to the buyers, a biotech company may license even its "crown jewels" to another company if it does not otherwise believe that it will have access to the financial resources to survive and to develop the product on its own. By analogy, car owners might well sell a perfect car, even in a discounted market, if they cannot afford to put gas in it.

Meanwhile, the party with plenty of cash for gas—that is, the pharmaceutical company—has an urgent need for the car. According to an estimate by Evaluate Pharma, the ten biggest drugs losing patent protection

---

[51] The exact percentages for each year can be found in the annual industry reports by Ernst & Young. Over the past couple of decades, each one of those percentages tends to hover between 20 and 30%, meaning that approximately one in two biotech companies operates with less than two years of cash.

in the 2007–2017 decade accounted for more than $915 billion in lifetime sales.[52] These lost revenues, combined with the research that shows that pharmaceutical companies have been more aggressive in acquiring biotech-derived products when their internal development efforts have been insufficient,[53] may explain why the value of biotech–pharmaceutical licensing transactions has been consistently increasing over the past few decades, reaching an all-time high in 2019 with $357 billion in deal value.[54] This need to fill the pharmaceutical pipeline is a strong enough motivation for the companies to pursue licensing transactions, despite the low likelihood that any of the products will succeed. This combination of factors has led to an extremely active biotech–pharmaceutical licensing market over the last decade. These factors are likely to be more significant influences on the likelihood of potential transactions than the possibility that any information imbalance exists in a fashion that would enable biotech companies to pawn off their less-promising products.

And so, it seems that economists may have described information asymmetry in biotech–pharmaceutical transactions largely because it was expected to be there, based on how economic theory applies in other fields. Viewing these transactions in light of the extremely low success rates found in the drug development business, however, suggests instead that the parties are more likely to be essentially symmetrical in their understanding of the *value* of the products being licensed, even if their knowledge of the technology itself differs. The above 90% failure rate suggests that the biotech–pharmaceutical licensing business is less like a market for lemons than for high-tech but unproven methods for buying lottery tickets. Almost all drug development projects will be the equivalent of Akerlof's totally defective car, or a losing lottery ticket, with no ability of either party to predict with confidence which product might turn out to be the one-in-ten or one-in-one-hundred opportunity that will be a successfully commercialized drug. Such symmetry under conditions of a likely product failure better explains the observed characteristics of the market—a large and robust market for licensing transactions, often with contingent payments but without

[52] See Eric Sagonowsky, "The Decade's Top 10 Patent Losses, Worth a Whopping $915B in Lifetime Sales," *Fierce Pharma*, August 17, 2017, https://www.fiercepharma.com/pharma/decade-s-top-10-patent-losses-featuring-seismic-sales-shifts.

[53] See Higgins and Rodriguez, "Outsourcing of R&D," 351.

[54] See Ernst & Young, *How Will Deals Done Now Deliver What the Health Ecosystem Needs Next? 2020 EY M&A Firepower Report* (London: Ernst & Young, 2020), 4, accessed online, https://assets.ey.com/content/dam/ey-sites/ey-com/en_gl/topics/life-sciences/life-sciences-pdfs/ey-firepower-report-2020-how-will-deals-done-now-deliver-what-the-health-ecosystem-needs-next-v2.pdf.

evidence of an accompanying lemons problem—than does a hypothesis of information asymmetry.

Symmetrical or not, the parties' relationship in alliances is clearly symbiotic. Each party needs what the other has in relative abundance, be it product candidates or hard cash. As a result, we are likely to continue to see a robust alliance market for as long as both biotech and pharma coexist, especially since, as we will see in Chapter 7, the biotech industry continues to be an extremely important source of new medicines.

6

The Biotech Entrepreneur

Who is the biotech entrepreneur? Is a biotech entrepreneur necessarily an accomplished scientist bent on improving human health by discovering a paradigm-shifting scientific insight? Some accounts of the industry certainly give that impression, and there have certainly been many biotech entrepreneurs who have been outstanding research scientists. One is Harvard Professor Walter Gilbert, who won the Nobel Prize in 1980 and served as CEO of Biogen, one of the several biotech companies he founded.[1] Another example is the long-serving founding CEO of the rational drug design company Vertex, Dr. Joshua Boger, a Harvard-trained chemist. He is portrayed in *The Billion Dollar Molecule* as the smartest student not only in his high school class but also "in the school, maybe in the school's history," the prodigy who wrote a four-hundred-page paper in the fourth grade.[2]

There is a tendency to imagine that since biotech is a science-based business, being a brilliant scientist is the principal prerequisite to be an entrepreneur in this industry. To be sure, the biotechnology industry has an unusually highly educated set of leaders, but the industry would be much smaller if a Walter Gilbert– or Joshua Boger–level of academic precociousness and achievement were required for leadership in the drug development business. Moreover, the CEO's task is far more extensive than managing scientific research projects, no matter how innovative or complicated they may be. A biotech company is an extremely complex business. It requires the application and coordination of scientific, regulatory, financial, and management expertise to operate effectively. Fundraising and communications capabilities, the ability to negotiate and manage business partnerships, problem-solving skills, strategic thinking, and perseverance are all essential for the company's success. So, while scientific knowledge is one critical aspect of building a successful biotech company, it is not the only one. This need for multidisciplinary leadership abilities helps

---

[1] See *Encyclopaedia Britannica Online*, s.v. "Walter Gilbert: American Biologist," last modified March 17, 2020, https://www.britannica.com/biography/Walter-Gilbert.

[2] Barry Werth, *The Billion Dollar Molecule: One Company's Quest for the Perfect Drug* (New York: Simon & Schuster, 1994), 142.

explains why, as the industry has developed, individuals with a variety of backgrounds have ended up with overall management responsibility.

Although biotech companies are not just science projects, they frequently start that way. Especially in the early years of the industry, companies' first CEOs were often the research scientists who invented the technology. They knew the core scientific principles and experiments better than anyone, and when the biotech business was new, there were no experienced "serial entrepreneurs" within the industry. Pharmaceutical company executives were often reluctant to leave highly profitable Fortune 500 companies to join nascent enterprises with what appeared to be dim prospects for success, so the research scientists became business executives. Some of these bench-to-business scientists took to the new tasks of capital formation, dealmaking, product development, and so on with enthusiasm and success; others preferred to return to the lab, either by stepping aside to take on a purely scientific role with the company or by returning to academia.

Because the industry continues to consist of several thousand companies, including numerous start-ups appearing every year, it is bound to include, at any particular time, a sizable portion of first-time entrepreneurs alongside others who are starting their second, third, or fourth company. In fact, according to a 2019 survey of biotech companies, the majority (56%) of newly appointed CEOs were first-timers.[3] So, where do all these entrepreneurs come from? What are their backgrounds and skill sets?

The Entrepreneurial Profile

Studies of entrepreneurs in various other industries have not necessarily linked advanced education with entry into a career in entrepreneurship,[4] and a high school diploma, or perhaps a short time in college, has clearly been

[3] See Karen Tkach Tuzman, "Rise of the First-Time Biotech CEO," *BioCentury*, December 19, 2019, https://www.biocentury.com/article/304098/the-management-crunch-in-biotech-is-creating-a-heyday-for-first-time-ceos. This is also consistent with reports from major VC funds that noted that the majority of their companies in their portfolio have "rookie" CEOs. See Bruce Booth, "Our Experience with First-Time Biotech CEOs: Five Behaviors That Matter," *Life Sci VC* (blog), May 25, 2016, https://lifescivc.com/2016/05/experience-first-time-biotech-ceos-five-behaviors-matter/, an article that also offers important advice to those new CEOs.

[4] Specifically, most studies have found no significant relationship between an individual's schooling level and the probability of becoming an entrepreneur. At the same time, for those who have decided to become entrepreneurs, there is evidence that a higher schooling level (that is, more years of education) has a positive effect on the performance of the venture. A thorough review of this literature can be found in Justin van der Sluis, Mirjam van Praag, and Wim Vijverberg, "Education and Entrepreneurship Selection and Performance: A Review of the Empirical Literature," *Journal of Economic Surveys* 22, no. 5 (December 2008): 795–841.

enough, even in high-tech fields.[5] Famous college dropouts Bill Gates, Steve Jobs, and Mark Zuckerberg are among the current era's most prominent technology entrepreneurs, with Steve Jobs declaring, in a Stanford University commencement speech (of all places), that dropping out of college was "one of the best decisions I ever made."[6] Well-known technology investor and entrepreneur Peter Thiel has even offered a $100,000 fellowship for people twenty-two years old or younger prepared to "skip or stop out of college" in order "to build new things."[7] In contrast, biotechnology may be able to lay claim to the title of having the most highly educated executives in any industry. To get a picture of just how highly educated they are, we compiled a summary of the backgrounds of the chief executives of public biotech companies listed on NASDAQ in the early 2000s.[8]

Of the approximately two hundred CEOs in our data set, 85% held an advanced degree, of which nearly two-thirds were doctorates.[9] An impressive 7% of the CEOs spent enough time in school to collect two doctorates, most often a combined MD/PhD. The biotech CEOs hailed from many colleges and universities, although it is interesting to note that 40% attended colleges dubbed by *Barron's* as among the country's "most competitive," and nearly 20% hold bachelor's degrees from the eight Ivy League universities. To put these educational accomplishments in perspective, a 2002 study of *Fortune's* seven hundred largest companies indicated that only 11% had earned Ivy League undergraduate degrees—just about half of the percentage of biotech Ivy Leaguers—and CEOs with PhDs tended to be a rarity in the large companies. This is consistent with a 2019 survey of *Fortune's* one hundred largest companies that found that less than 5% of their CEOs held PhDs.[10] Even among the major pharmaceutical companies, not even one-third of the CEOs held an MD or PhD degree, with the MBA being the degree of choice

---

[5] That is not to mention studies suggesting a link between juvenile delinquency and subsequent entrepreneurship. See, for example, Martin Obschonka et al., "Rule-Breaking, Crime, and Entrepreneurship: A Replication and Extension Study with 37-Year Longitudinal Data," *Journal of Vocational Behavior* 83, no. 3 (December 2013): 386–96.

[6] H. O. Maycotte, "Education vs. Entrepreneurship: Which Path Wins?," *Forbes*, June 2, 2015, https://www.forbes.com/sites/homaycotte/2015/06/02/education-vs-entrepreneurship-which-path-wins/#2c3660714cdc.

[7] Thiel Fellowship (website), accessed September 14, 2020, http://thielfellowship.org/.

[8] Information about the CEOs of publicly traded biotech companies was derived from a review of the proxy statements of biotech companies filed with the Securities and Exchange Commission in 2003.

[9] The unusually high number of PhD-holding founders in the biotech industry is also noted in Waverly W. Ding, "The Impact of Founders' Professional-Education Background on the Adoption of Open Science by For-Profit Biotechnology Firms," *Management Science* 57, no. 2 (February 2011): 259, who finds that among the 1,090 biotech founders in her data, 49.5% had a PhD.

[10] See Kimberly A. Whitler, "New Fortune 100 CEO Study: The Top Graduate Schools Attended by Fortune 100 CEOs," *Forbes*, September 28, 2019, https://www.forbes.com/sites/kimberlywhitler/2019/09/28/new-fortune-100-ceo-study-the-top-graduate-schools-attended-by-fortune-100-ceos/#7947f67a2719.

for almost all of the others. There is little doubt that biotech chief executives are not only a very bright and ambitious group but also one that has put a premium on advanced levels of education.

By the time the companies reached the initial public offering (IPO) stage—and were therefore included in our survey—not quite 10% of the CEOs had entered the industry directly from academia, although we believe that the percentage of academics-as-CEOs would be considerably higher in a survey of younger, still-private companies. Other CEOs of these publicly traded biotech companies came from a variety of backgrounds, from Wall Street to private medical practice, but the vast majority, about 75%, most recently had worked in either the pharmaceutical industry or other biotech companies. Table 6.1 shows the industries in which biotech CEOs worked immediately prior to their current positions.

These data about the educational and professional backgrounds of biotech CEOs may point most clearly to the conclusion that there is a range of pathways to biotech senior leadership. This phenomenon can be seen in the answers several successful industry leaders gave when a recent college graduate asked what she should do to launch a career as a biotech entrepreneur. The former research scientist said, "Get a PhD in molecular biology." The deal-making lawyer said, "Go to law school," while the business development executive said, "Business school."

The consistent theme that appears in these comments, as well as in our survey data, is the potential value of higher education, or at least of having one or more advanced degrees. None of those degrees, however, provides in-depth professional preparation for all of the aspects of a career combining scientific discovery with entrepreneurial business building in a highly regulated environment in which the company's primary assets are patents and cash. Since few, if any, potential biotech entrepreneurs will have all of the relevant

Table 6.1 Backgrounds of biotech CEOs

Industry	Percentage of CEOs
Biotech	41
Pharmaceutical	34
Academia	9
Finance	7
Medicine	2
Other*	7

*Other includes a wide range of industries, from communications and electronics to chemicals and defense.

degrees, the postgraduate degree(s) they do hold probably signal two related characteristics. The first is advanced knowledge and expertise in at least one (or perhaps two) of the various fields relevant to the biotechnology business. The second is a demonstrated interest in learning and a well-trained ability to do so.

The first characteristic, a deep knowledge base in a relevant field, is likely to be most important for initially getting a foothold in the biotech industry. Few companies have the luxury of hiring people who cannot make an immediate hands-on contribution to an important task. Only the largest biotech companies are able to bear the expense of following the large pharmaceutical companies' lead in recruiting newly minted MBAs into jobs that position the employee for the kind of career growth that can end up with an office in the executive suite. Such career paths in large companies often involve opportunities not only to take on increasing levels of responsibility in the employee's initial area of expertise but also to become exposed to enough other areas that they obtain a good understanding of the entire organization. This process helps prepare them for taking on senior leadership positions. In the large universe of small biotech companies, however, researchers, physicians, lawyers, and other professionals will usually be hired to perform the specific jobs for which they have been highly trained, both at universities and in prior jobs. The company's goal is to find someone who can make the immediate substantive contributions it needs. Those employees who end up advancing to more senior business roles typically do so because they have made good use of the second characteristic of all that education—knowing how to learn complicated new things quickly.

Are You an Entrepreneur?

Holding an advanced degree may be a common characteristic of biotech executives, but by itself it does not automatically identify potential entrepreneurs. Due to the complexity of the industry and the range of skills required, even creating a new product opportunity or technology may not signal that someone would thrive in an entrepreneurial setting. While there are many guides offering advice on being a successful entrepreneur, there are fewer resources available to help potential entrepreneurs think about whether it is the right career path for them. In other words, "How do I know if I'm an entrepreneur?"

To shed further light on the qualities and traits that are helpful for biotech entrepreneurship, we have developed a set of questions to help people consider whether they might find a career in biotechnology entrepreneurship fulfilling and a good match for their personality and inclinations.[11] There may be other important considerations, but these represent a good place to start, and they are basically an entrepreneurial version of the Delphic Oracle's injunction to "know yourself."

(1) *Do you always think there is a better way to do things?* When biotech CEOs are not working to raise money, they often describe their jobs as creative problem-solving. Their companies are trying to succeed in the same task being pursued by the research divisions of large pharmaceutical companies, but with a tiny fraction of the resources. Creativity is essential, and since the industry is subject to countless laws and regulations, that creative spirit needs to be channeled into pathways that will work effectively in a very carefully regulated environment.[12]

(2) *Are you willing to take on just about anything, even if you don't know much about it?* In a nearly constant state of having insufficient resources, biotech companies cannot afford to have staffs of experts in every possible area. Entrepreneurial employees need to banish the phrase "That's not my job" and replace it with enthusiasm for the opportunity to learn—and master—something completely new and difficult whenever necessary.[13]

---

[11] The development of this approach was led by Dr. Lisa Drakeman in countless conversations with MBA and PhD students and others considering entering the industry. See, for example, "Dr. Lisa Drakeman, Former CEO of Genmab: Entrepreneurs Must Accept Rejection but Keep Going," News & Insight, Cambridge Judge Business School, last modified June 18, 2013, https://www.jbs.cam.ac.uk/insight/2013/dr-lisa-drakeman-former-ceo-of-genmab-entrepreneurs-must-accept-rejection-but-keep-going/.

[12] A number of studies find that entrepreneurs are characterized by higher levels of achievement motivation (need for excellence and accomplishment of ambitious goals). Specifically, they find that entrepreneurs score significantly higher in the achievement facet compared with managers and other professionals. See, for example, Hao Zhao and Scott E. Seibert, "The Big Five Personality Dimensions and Entrepreneurial States: A Meta-analytical Review," *Journal of Applied Psychology* 91, no. 2 (March 2006): 259–71; Christopher J. Collins, Paul J. Hanges, and Edwin A. Locke, "The Relationship of Achievement Motivation to Entrepreneurial Behavior: A Meta-analysis," *Human Performance* 17, no. 1 (2004): 95–117; and Wayne H. Stewart Jr. and Philip L. Roth, "A Meta-analysis of Achievement Motivation Differences between Entrepreneurs and Managers," *Journal of Small Business Management* 45, no. 4 (October 2007): 401–21.

[13] Research shows that a high degree of self-efficacy (a person's beliefs that they can perform tasks and fulfill multiple roles) increases the likelihood of both being a nascent entrepreneur and creating an operating business. See Gavin Cassar and Henry Friedman, "Does Self-Efficacy Affect Entrepreneurial Investment?," *Strategic Entrepreneurship Journal* 3, no. 3 (September 2009): 241–60. Many an entrepreneur has also been found to be a jack-of-all-trades, capable of performing a variety of tasks rather than highly specialized ones. See, for example, Edward P. Lazear, "Balanced Skills and Entrepreneurship," *American Economic Review* 94, no. 2 (May 2004): 208–11; Edward P. Lazear, "Entrepreneurship," *Journal of Labor Economics* 23, no. 4 (October 2005): 649–80; J. Wagner, "Testing Lazear's Jack-of-All-Trades View of Entrepreneurship with German Micro Data," *Applied Economics Letters* 10, no. 11 (2003): 687–89; Joachim Wagner, "Are Nascent Entrepreneurs 'Jacks-of-All-Trades'? A Test of Lazear's Theory of Entrepreneurship with German Data," *Applied Economics* 38, no. 20 (2006): 2415–19; and Olmo Silva, "The Jack-of-All-Trades Entrepreneur: Innate Talent or Acquired Skill?," *Economics Letters* 97, no. 2 (November 2007): 118–23.

(3) *Are you comfortable taking risks?* Success is not along a defined path, and obstacles are the norm. Many highly credentialed graduates have succeeded in their academic work by having a detailed and comprehensive understanding of a particular field, and they are used to seeking and finding the "right" answers. That is not always—in fact, not often—possible in biotech company leadership. Plans need to be made and often changed without adequate amounts of information being available, and many of those plans will go wrong, usually in unexpected ways. For some, that opportunity to juggle many things in unpredictable environments sounds exhilarating; for others, it would be a highly unattractive way to devote countless hours each week.[14]

(4) *Do you like to do new things, or do you prefer routine?* As with all of the earlier questions, there is no right or wrong answer, just the one that suits an individual's preferences. It is not uncommon for biotech senior executives to find that their day goes in completely unexpected ways on a regular basis. Whether the surprises are good (the stock jumps up in a way that may lead to a new financing opportunity) or not (the FDA calls to place a "clinical hold"—that is, stop patient enrollment—on your most important product in clinical trials), there are a lot of them.[15]

(5) *Can you accept rejection and failure?* Perseverance is a must in an industry with a 90% failure rate and a financing environment in which most potential investors (sometimes all investors) say "no" most of the time. Multi-hundred-million-dollar corporate alliances can, and often will, fall apart at the last minute after a year or longer spent in detailed negotiations. Being able to plot a course toward success is important, but it is just as essential to be able to figure how to bounce back from disappointment or failure.[16]

---

[14] The connection between entrepreneurship and risk dates back to Frank H. Knight, *Risk, Uncertainty, and Profit* (Boston: Houghton Mifflin, 1921), who argued that entrepreneurs are characterized by their ability to take risks despite significant uncertainty. Formal mathematical models have also shown that, in equilibrium, more risk-averse individuals become workers while the less risk-averse become entrepreneurs. See Richard E. Kihlstrom and Jean-Jacques Laffont, "A General Equilibrium Entrepreneurial Theory of Firm Formation Based on Risk Aversion," *Journal of Political Economy* 87, no. 4 (August 1979): 719. More recent studies have also found empirical support for the relationship between risk aversion and the likelihood of becoming an entrepreneur. See, for example, Lazear, "Entrepreneurship," 649–80; and Robert E. Hall and Susan E. Woodward, "The Burden of the Nondiversifiable Risk of Entrepreneurship," *American Economic Review* 100, no. 3 (June 2010): 1163–94.

[15] Studies comparing the personality traits of entrepreneurs versus managers have consistently shown that entrepreneurs are more open to experience than managers. See Zhao and Seibert, "Big Five," 259. Openness to experience (one of the Big Five personality traits) describes the breadth, depth, originality, and complexity of an individual's mental and experimental life. See Zhao and Seibert, "Big Five," 261.

[16] Tolerance for failure is closely related to the concept of risk aversion in the studies mentioned earlier. See notes 14–15. A recent study has also found explicit support that entrepreneurs display the greatest tolerance for risk compared with nonfounder CEOs and inventor employees. See Robert E. Hall and Susan E.

There is no guaranteed formula for success as a biotech entrepreneur. However, potential biotech entrepreneurs would probably do well to be able to answer "yes" to these questions, and, in fact, entrepreneurs in almost any field should probably be able to answer in the affirmative to most of them.[17] Novel ideas, creative thinking, clever new approaches—the stock-in-trade of entrepreneurs everywhere—are an essential component of success. But at the same time, all of that creativity needs to be exercised and managed in ways that will gain the support of the investors and regulatory bodies that are crucial to the company's success. For example, carefully maintaining meticulous, detailed written records in full compliance with the impressively complex and comprehensive regulations established by the FDA and other similar bodies around the world is as important for successfully launching a new medicine as creating the product in the first instance. Managing this delicate balance in a constantly changing environment requires an exceptional level of adaptability as well as the ability to learn new things quickly. While no specific academic discipline can claim ownership of this skill, it is arguably one of the main benefits of advanced education. This could help explain the unusually high levels of education of the biotech leaders and the striking diversity of the educational and professional backgrounds of those leaders. Whatever their backgrounds may be, those who enjoy the type of environment we have described and who are able to tolerate operating in a risk-laden environment of complexities and uncertainties can find biotech entrepreneurship to be an exciting career path. They will have the rewards of working with a group of incredibly smart people who single-mindedly focus on building a successful business by bringing important new medicines to the patients who need them.

As we conclude this section on entrepreneurial characteristics, it is worth noting that there is very little research in this area and that developing a better understanding of the relationship between personality traits and successful entrepreneurship is an important area for future study. To the best of our knowledge, there are no studies that specifically examine such connections in the context of the biotech industry. Given the variety of the different tasks and stakeholders involved, it would be interesting for further studies to shed light on how certain qualities and traits (or lack thereof) affect the processes and

---

Woodward, "The Burden of the Nondiversifiable Risk of Entrepreneurship," *American Economic Review* 100, no. 3 (June 2010): 1163–94.

[17] For a more general discussion regarding whether individuals should consider working in an entrepreneurial environment, see Jeffrey Bussgang, "Are You Suited for a Start-Up?," *Harvard Business Review* 95, no. 6 (November/December 2017): 150–53. For a thorough review on the personality traits of entrepreneurs, see also Sari Pekkala Kerr, William R. Kerr, and Tina Xu, "Personality Traits of Entrepreneurs: A Review of Recent Literature," *Foundations and Trends in Entrepreneurship* 14, no. 3 (July 2018): 279–356.

outcomes of those interactions. For example, are some traits more effective in managing the company's own board or shareholders versus negotiations with a potential licensee such as a large pharmaceutical company, or vice versa? As discussed in the references in the notes to this chapter, the vast majority of studies on the personality traits of entrepreneurs do not account for industry characteristics, and the performance measures are quite generic, if they are captured at all.

Setting Off on the (Sometimes Turbulent) Journey

As in any industry, biotech CEOs face core challenges. They need to establish the company's goals, set a course that allows the company to achieve those goals, marshal the resources necessary to follow that course, and employ those resources as efficiently as possible. We are not sure whether there is any industry in which the CEO can say "Make it so," and legions of devoted employees immediately dedicate themselves to delivering successful results on time and under budget. Biotech is certainly not it.

Biotech entrepreneurs need to contend with a remarkable degree of complexity in dealing with both the variety of the company's tasks and the uncertainty of the environment. Even if they have a potential cancer cure in the lab, the scientific discovery is only the beginning of the long, expensive, and heavily regulated process of developing a new therapeutic product. Successfully completing that process requires entrepreneurs to work with a diverse array of essential constituencies, not all of whom always have fully compatible needs and interests. That is, beyond the core challenge of finding a new and important way to treat a life-threatening or seriously debilitating disease—and, in fact, just to have the chance to try to accomplish that goal— biotech entrepreneurs will find that they need to work closely with a range of different types of stakeholders. In this section, we take a closer look at what the constituencies are and why they are influential. We also offer some suggestions about how biotech entrepreneurs can manage those relationships.

We will describe five key stakeholders. These include investors; employees and other human resources; corporate partners; physicians, patients, and payers; and the government agencies that regulate the industry and approve new medicines.[18] One of the key reasons for identifying these influential

[18] Others could probably offer excellent suggestions for expanding the list. Therefore, we do not mean for this to be an exhaustive catalog, but it can serve as a useful approach to explain the layers of activities and interests that are involved in moving a biotech company in the direction of developing and commercializing a new medicine.

and important groups is to emphasize the need for entrepreneurs to be able to speak convincingly to each in their own language. As entrepreneurs deal with these groups, they will rarely, if ever, have the upper hand. Accordingly, they will need to manage the relationships primarily by articulating persuasive arguments showing that if the stakeholders can help the company fulfill its mission, the stakeholders will, at the same time, be successfully achieving their own goals.

Investors and the Fundraising Process

A central responsibility of biotech entrepreneurs is to acquire the financial resources necessary for the company's ongoing operations. Biotech is a capital-intensive industry with long development timelines and years of loss-making operations. Unlike the major pharmaceutical companies, which have the ability to fund their R&D efforts with revenues from the sales of existing products, the vast majority of biotech companies are working on their first product. Accordingly, because it costs so much to develop a new medicine, including the nearly inevitable false leads, delays, and failures along the way, biotech companies need continued access to investment capital. Moreover, since their major assets are intangible ones, primarily intellectual property, there are few opportunities to borrow significant amounts of money, as might be possible in industries with more easily financeable real estate or heavy equipment. As a result, and as we discussed extensively in Chapter 4, biotech companies are constantly working to attract new investors and to maintain good relations with the existing investors who might consider increasing their investment in the future.

Unlike the entrepreneurial ventures in other industries, which can make meaningful progress from the cash generated by the founders "maxing out" their credit cards and taking out second mortgages on their houses, biotech companies typically need to access tens or even hundreds of millions of dollars in investment capital. Therefore, the ability to convince venture capitalists and other potential investors to invest in the company is an essential part of the job of a biotech entrepreneur. To get the cash needed to pay for the company's activities, entrepreneurs must convince potential investors that the science is sound, the development plans are sensible, and, *crucially*, that investors will make a very good financial return. In most biotech companies, the process of doing so will occupy a very large amount of the entrepreneur's time.

Being able to tell a clear and compelling story to venture capitalists and other potential investors is, therefore, an important skill for all biotechnology entrepreneurs. This is a task not only for the CEO but also for every member of the management team. When it comes to raising money or forging alliances, managers responsible for finance, research, business development, and other areas need to be able to talk knowledgeably about all aspects of the company. Even more importantly, having a shared understanding of what the company is doing, why it is doing it, and where it is going can not only present a consistent story to potential investors but also help keep the team focused on the same overall goals.

The initial presentation will sometimes be made to a single member of a venture capital firm. At other times, it may take place at a "venture fair" or "entrepreneurial showcase" with a large audience. In either case, the goal is to generate sufficient interest in the company so that potential investors would be willing to spend more time learning about it. Any financing transaction will take place only after a series of increasingly detail-oriented meetings, including the lengthy due diligence process. Therefore, the first presentation does not need to contain as much information about the company as you can squeeze into the allocated time. Instead, it needs to motivate the people in the audience to put your company on the very short list of investment possibilities that they should consider seriously enough to be willing to devote the time for more meetings. To do so, it should focus on why your company can create very significant financial value by succeeding scientifically.

The standard version of the presentation should aim for a duration of about twenty to thirty minutes and will be accompanied by about twenty to thirty PowerPoint slides. (These numbers are just general guidelines but are at least a good place to start.) In general, the presentation will include information about who the key people are, what the core technology or product is, and what important milestones are on the horizon. In considering which milestones are the most important, think about what technological or other achievements could signal to these investors—and to future investors, partners, or acquirers—that the company has become considerably more valuable because they have been achieved. In other words, the presentation will show that, as progress is made, the risks diminish and the opportunities grow.

As much as individual venture capitalists may find your technology scientifically exciting, they will ultimately choose those companies where they think they will obtain the best financial returns at the exit—that is, when they sell the company or sell their shares of stock after a successful IPO. They will definitely focus on detailed questions of science and medicine, especially in those subsequent meetings, and, in fact, many will have PhDs in the sciences or MDs. It is critical to remember, however, that they are employed to manage investment funds. Their responsibilities and goals are to make money for their investors. They will be deciding not just whether they believe in the potential of the company's technology or product but whether future acquirers or IPO investors will become believers as well, especially after the key milestones have been achieved.[19]

In these initial presentations, entrepreneurs need to aim for "clear and convincing," not "detailed and comprehensive." Venture capitalists will hear quite a few other biotech company presentations that week—probably even that day—and your presentation is competing for their attention with all of the other molecules and medicines. Moreover, one thing investors will be looking for is whether the management team can make a good presentation. If they ultimately decide to invest, they will need you to help them convince other members of the venture capital syndicate, and, as progress is made, potential acquirers or IPO investors in the future. So, the opportunity to make a clear, crisp, and compelling first presentation is essentially an audition as well as a chance to convey information about the company.

Because investors are constantly evaluating opportunities throughout the industry, the presentation should ideally be crafted with an eye toward where this company is situated. In some cases, a biotech company is breaking ground in an area so new that there is nothing comparable throughout the entire universe of 5,000+ companies. More often, the company's focus will be to develop a novel approach to an existing technology or to apply a well-known technology in a new way. Knowing where your technology or product fits in the universe of biotech companies will provide a helpful way to shape the presentation. Showing where the company fits within the industry and why there is a place for it can also help the individuals hearing your presentation talk to their colleagues at the firm about the company.

One of your key tasks will be to give the audience a "hook"—a shorthand way to remember and describe the company. Bear in mind that your

---

[19] First-time entrepreneurs, especially those coming from academia, may be surprised, or even dismayed, by the amount of time we spend in this book—and the amount of time they will spend as entrepreneurs—talking about financial topics rather than scientific ones. "Why are you talking about money so much?" they have asked us at various entrepreneurial events. The short answer is that if you have a great scientific insight into how to solve a medical problem, the only way to make that solution available to the patients who need it is to spend the massive amount of money required to reach FDA approval. The most likely sources of that money are investors who earn their livings by "buying low and selling high." The most compelling biotech stories are the ones that show investors how achieving your goals will help them achieve theirs. Biotech is not unique in this regard, and one of us has offered the same suggestion for those seeking funding for the humanities: "A rule of thumb frequently employed in fundraising—whether in the not-for-profit context or the search for investment capital—is to focus the conversation on the interests of the person holding the checkbook. It is easy to see that things will be better for the people who are, in the fundraising vernacular, 'making the ask.' Their university or soup kitchen or start-up will have more funds with which to achieve its goals. The key question for the grant-giver or investor is, 'Why will funding this particular organization, out of all of the many excellent requests I have received, best fulfill my mission?'" (Donald Drakeman, *Why We Need the Humanities: Life Science, Law and the Common Good* [New York: Palgrave Macmillan, 2016], i).

overriding goal is to secure the next meeting. For that to happen, the venture capitalists who heard your pitch now need to convince their colleagues that it would be worthwhile for the firm to dig deeper into the possibility of investing in the company. Each venture capital firm has its own process for deciding which companies deserve a second look. A fairly typical one involves regularly scheduled meetings of the firm's partners during which all the new opportunities of the past week (or month) will be discussed. That could involve twenty or more companies discussed in an hour, with zero to perhaps 10% of them offered a chance for that next meeting.

What should the venture capitalist who saw your presentation say about the company in the roughly sixty seconds or so available in the team meeting? Whatever it is, it should be in bullet point form within the first three slides—ideally, summarized on the first and last slides of the presentation. These time-stressed investors are likely to remember only one key thing from your presentation—maybe two or three at most. In preparing the talk, you will need to spend a great deal of time thinking about what that key "take home" message should be.

The best way to get a sense of how to craft that message is to get advice from people who know the industry well. They could be the company's advisors or board members, or perhaps industry executives or venture capitalists you come across at meetings of local or national industry organizations.[20] As an investor-turned-entrepreneur told the audience at one such meeting, venture capitalists are often very stingy with their cash but can be extremely generous with their advice. It can be helpful, therefore, to start talking to them before a fundraising process starts to get that advice—and to begin to build a relationship that will help when it comes time to raise money.[21]

Once investors have become interested enough to provide funding to a company, entrepreneurs will quickly learn how much of a role those shareholders, and the board members whom they have designated, will play in the way the company evaluates strategic and operational alternatives. One of the central jobs of the management team is to make decisions that will increase shareholder value, but views on exactly how best to do so may vary not only among the individuals who are part of that team but also among the board members

---

[20] Such meetings not only offer networking opportunities but may also provide very useful guidance on just this sort of thing. See, for example, Biotechnology Industry Organization, "How to Pitch to Investors," October 6, 2014, https://www.bio.org/blogs/how-pitch-investors.

[21] The same principles typically apply to other communications efforts by the company, such as press releases. There can be a tendency in science-based companies to compose announcements describing scientific advancements in detailed scientific terms. While it is important to get the science right, it is equally important to bear in mind that the audience will be primarily composed of the media, investors, and the public. Too much scientific language could dilute the basic message that important progress has been made.

and shareholders. Accordingly, major corporate decisions will be made at least partially in the context of how the CEO and the board expect new or existing investors to react and how that reaction will affect the company's ability to raise funds in the future.

For venture capital–backed private companies, the venture capitalists (who typically occupy all, or at least most, of the seats on the board of directors) will ultimately have either an actual or a de facto financial veto on development plans. Venture investments typically come in "rounds," normally labeled alphabetically, that are designed to provide just enough cash for the company to reach a predetermined milestone,[22] and those rounds are often broken into increments known as "tranches" so that the cash will be released to the company only when specific signs of progress have been achieved. Venture financings often involve a syndicate composed of several venture firms, and biotech CEOs may find that the various venture capitalists serving on the board of directors do not always agree with one another about the right next steps, with the CEO responsible for figuring out how to obtain the necessary alignment.[23] This can be a difficult and delicate task in an environment in which venture capitalists are rarely reluctant to replace CEOs.[24]

Successfully negotiating the venture capital stages can open up the possibility of doing an IPO. An IPO offers an opportunity for the company to raise additional funds while providing the venture capital investors with an opportunity to sell their shares on the public market. The IPO typically broadens the shareholder base to include many new types of investors. Shareholders of public companies are generally a much more diverse group than life science venture capitalists, and their interests vary accordingly. This presents some additional challenges for biotech executives' communication skills. For example, while all shareholders want the stock to become more valuable, some investors will be content with (and anxious for) a fairly quick profit, while others have the patience to wait for much larger increases in value over a longer period. Long-term investors are typically prepared for the company's stock to rise and fall over time, as they wait for a transformational event that will dramatically increase the company's value, such as successful Phase III clinical trials, FDA approval, or perhaps an acquisition by a major pharmaceutical

[22] See Paul A. Gompers, "Optimal Investment, Monitoring, and the Staging of Venture Capital," *Journal of Finance* 50, no. 5 (December 1995): 1464–65; and Chapter 4.

[23] For example, research has shown that young venture capital firms take companies public earlier than older venture capital firms in order to establish a reputation and successfully raise capital for new funds. See Paul A. Gompers, "Grandstanding in the Venture Capital Industry," *Journal of Financial Economics* 42, no. 1 (September 1996): 133.

[24] For a discussion on CEO replacement by VCs, see Thomas Hellmann and Manju Puri, "Venture Capital and the Professionalization of Start-Up Firms: Empirical Evidence," *Journal of Finance* 57, no. 1 (February 2002): 169–97, and references therein.

company. Meanwhile, shareholders with a much shorter attention span may be interested only in seeing the stock "pop" as soon as possible, giving them a quick profit when they sell their shares to new investors. As a result, these different shareholders may have conflicting views of some potential corporate events.

Consider how these two different types of shareholders would think about a company's decision to out-license its most advanced (that is, closest to FDA approval) product in return for a large upfront payment but only modest royalties on future sales. Short-term investors would be likely to support the deal since the announcement—"Biotech Company receives $150 million upfront payment from Major Pharma Company"—could provoke a jump in the stock price, thus providing an opportunity for a quick profit. Meanwhile, long-term investors could object to the out-licensing deal because for them, selling the "crown jewel" without a chance to receive a significant share of future profits has, in fact, diminished the ultimate future value of the company, even if the stock responds positively in the short term. They would rather have a tenfold profit in the future than a 10%–20% "uptick" in the stock this week.

The ideal solution would be to find a partner willing to give the company a large upfront payment plus the opportunity for a substantial share of future sales and profits. These kinds of terms are not easy to obtain. They occur most often when the biotech company has an unusually strong negotiating position, which may require impressive data in clinical trials. Long-term investors may be prepared to wait for late-stage clinical data to arrive. Yet even if the company is able to raise new funds at reasonable prices (reasonable at least in the eyes of management and the board of directors), short-term investors may be unhappy that they have to wait what seems to them to be an unreasonably long time for the extensive clinical trials to be completed. The result may be pressure from short-term investors to consider an alliance sooner rather than later. In short, it is extremely challenging for biotech management to try to please all of the investors all of the time, even when the stock is going up. (When the stock is going down, it can seem considerably easier to displease all of the investors at the same time.)

Employees—and the Rise of the Virtual Biotech Company

A key strategic and operational question for any new biotech company is: Do we need any employees, and if so, which ones? With the dramatic growth of the drug development outsourcing business, employees have increasingly become an option rather than a necessity. There are significant trade-offs involved in making the decision to implement the company's plans primarily with in-house resources (where the entrepreneur has more control) or via outsourcing (which offers financial flexibility with the likelihood of less control). Outsourcing some of the work of developing a new medicine is inevitable because of the impressive variety of expertise needed, as well as the human resources required to run late-stage clinical trials. Even the largest pharmaceutical companies regularly retain contract research organizations (CROs). The question for the entrepreneur is whether it would be better to have full-time, dedicated employees.

This concept of a predominantly virtual biotech company has developed over time. In biotech's early days, it was not uncommon for biotech companies to build their own laboratories in impressive-looking buildings. In those labs, companies could hire and train experts in the new technologies being employed and would have all R&D efforts under their full control. Over time, as the amount of funding available to the industry did not always expand as fast as the number of companies did, and as many companies were created around specific products rather than core technologies, well-equipped labs full of highly paid employees were increasingly seen as expensive and not necessarily essential. As discussed in earlier chapters, a large and highly experienced contract research industry grew alongside the biotech industry, and it is capable of performing virtually all of the tasks involved in drug development, from identifying candidate molecules at the earliest stages of discovery research all the way through selling the FDA-approved product.[25] A CEO with R&D or project management experience can, in fact, be the sole full-time employee.[26] These lean biotech structures are a considerable contrast with the large pharmaceutical company R&D organizations, which not only employ thousands of researchers—the top twelve global pharmaceutical companies alone employ a total of approximately 890,000 employees worldwide[27]—but also retain the services of many of the same CROs used by biotech companies.

The CROs can essentially do it all, including discovery, development, manufacturing, clinical trials, and regulatory submissions, not to mention

---

[25] For a detailed analysis of the CRO industry, see Erin Wilson, Robert Willoughby, and Mark Wallach, *CRO Industry Primer* (Zürich, Switzerland: Credit Suisse, 2016), https://research-doc.credit-suisse.com/docView?language=ENG&format=PDF&source_id=csplusresearchcp&document_id=807217850&serialid=bWxS47N1G4l%2f5p8umJGkcSVAYjNNuJYDxdjMq6OmgRA%3d.

[26] See, for example, Jeanne Whalen, "Virtual Biotechs: No Lab Space, Few Employees," *Wall Street Journal*, June 4, 2014, https://www.wsj.com/articles/virtual-biotechs-no-lab-space-few-employees-1401816867.

[27] See Stephen Naylor and Kirkwood A. Pritchard Jr., "The Reality of Virtual Pharmaceutical Companies," *Drug Discovery World* (Summer 2019): 9, https://www.ddw-online.com/business/p323009-the-reality-of-virtual-pharmaceutical-companies.html.

marketing, sales, accounting, legal work, and virtually every other corporate function. If a biotech company really is, as we have suggested, a marriage of money and molecules, there are only two essential jobs: someone to raise the money (usually the CEO's primary role) and someone to turn the molecules into medicines, which can be done internally, as often used to be the case, or externally, as is becoming increasingly possible. Virtual or "lean" biotech companies can reduce the cost of drug development by paying for what they need only when they need it. Even with a CRO's profit markup on the services it provides, having just-in-time access to specific expertise in light of the breadth of a biotech company's research and development needs can make this a cost-effective and efficient approach to drug development.

The fact that companies can make significant R&D progress with very few employees should not be a sign that only the CEO is an essential element of success. Quite to the contrary, the individuals doing the day-to-day work are vital resources. As veteran biotech venture capitalist Bruce Booth has noted, "Human capital is a company's most important asset."[28] But those critical assets are not necessarily employees of the company. They may be university researchers, members of a CRO's staff, independent consultants, scientific advisory board members, and so on. To be able to access the most talented people on a just-when-you-need-them basis is a financial luxury that is accompanied by a managerial challenge.

Managing a workforce of extremely smart, well-educated, scientifically trained individuals requires a great deal of clarity in communicating the company's goals and in providing the right kinds of incentives to keep those valuable human resources aligned with those goals. Those communication and incentive issues become even more complex when many of the key individuals are not employees of the company. The lean companies' managers can only indirectly control the day-to-day drug development activities, and it can sometimes be a very significant challenge for small biotech companies to get the attention of the CROs, which often have considerably larger agreements with major pharmaceutical companies. As a result, biotech entrepreneurs may need to dedicate significant time and resources to identifying the right CROs, negotiating a mutually beneficial contract with them, and then managing that relationship through the course of the work under the contract.

Relying almost entirely on CROs is a potentially viable approach for some biotech companies but is certainly not without complications and risk. As a result, many biotech companies opt to become less lean by adopting a hybrid model in which entrepreneurs set out to create the most appropriate blend of in-house and CRO work. In whatever manner executives and the board of directors ultimately decide to divide tasks between employees and CROs, the use of CROs has become virtually essential in a business requiring expertise in so many areas, not to mention the need to have products manufactured for clinical trials in FDA-compliant facilities that are often too expensive (and time consuming) for biotech companies to build from scratch. Therefore, finding a way to manage a mix of in-house and outside resources is a significant and essential challenge for virtually all biotech CEOs.[29]

These "virtual" biotech companies have increasingly attracted the interest of venture capitalists (VCs)—according to an estimate of one experienced venture capitalist, nearly a third of the annual $4 billion in U.S. venture funding goes to such companies. They also have been able to recruit prominent biotech leaders to serve on their boards, including the former CEO of Vertex, Joshua Boger, who was drawn by the "elegance of the science" and the opportunity to build a company cheaply, rather than "helping someone hire all their vice presidents."[30] Nevertheless, there are numerous examples of biotech companies with only one or perhaps a handful of employees that have raised—and spent—millions of dollars on external research, with good enough results to be acquired for substantial sums by large pharmaceutical companies.

An especially promising area for future research pertains to the increasingly popular model of the virtual biotech. To this date, our understanding of the pros and cons of this model is primarily based on anecdotal evidence. For example, one vital aspect of the virtual biotech model is managing the relationship with CROs, yet there are no studies that discuss such relationships. A rigorous series of studies would allow managers and management scholars to develop a framework that could inform entrepreneurs, investors, and other stakeholders about the conditions that favor such virtual models versus alternative ones.

Corporate Partners

Biotech companies need regular infusions of cash to continue their development efforts, and along with private and public investors, alliance partners provide the greatest amount of cash to the biotech industry. Due to the long

---

[29] See Heidi Ledford, "Virtual Reality," *Nature* 498, no. 7452 (June 2013): 127–29.
[30] Quoted in Whalen, "Virtual Biotechs."

development timelines and high cost of drug development, biotech companies can be "development-stage" entities, having not yet successfully brought a product to the commercial market for a decade or more—in some cases, for many decades. Because biotech companies have no recurring sources of revenues from product sales to fund their R&D activities, current or potential alliance partners represent another critical constituency, and entrepreneurs are likely to devote a considerable amount of their time courting potential partners. Once that courtship has yielded an alliance, an equally large amount of time will be needed to manage that relationship.

As discussed at length in Chapter 5, corporate partners typically acquire rights to future products rather than buy finished products, and, as a result, they often have a significant influence on the biotech company's day-to-day operations, especially when (as is frequently the case) the product around which the alliance has been created is the biotech company's primary asset.

A pharmaceutical partner will typically insist on having a strong, or even dominant, role in all key aspects of the product's future development. It has negotiated for that influential position by insisting on having a collection of control rights in the alliance agreement. In fact, even when that future development effort is described as a 50–50 partnership of equals, the pharmaceutical company will often insist on having the "casting vote"—that is, in the event the companies cannot reach agreement on any particular decision, the pharmaceutical company ultimately has the right to set the future course. The degree to which the partner holds these critical control rights can therefore dramatically affect both the nature and the pace of future development activities. It is not unusual for large pharmaceutical partners to favor more careful, stepwise decision-making based on larger amounts of data than the high risk–high reward strategy often favored by entrepreneurial companies and their investors, who often view the "big pharma" style of development as one that inevitably takes longer, costs more, and yields fewer successes.

Biotech companies also learn that there are usually multiple layers of committees required for decision-making inside the large companies, a process that may require substantially more time to reach an authoritative decision than the biotech companies expect, even on matters where both companies are fully aligned. For the biotech company, the partnered product may be the crown jewel, but within a large pharmaceutical company that is managing dozens or even hundreds of ongoing R&D projects, many decisions about this particular alliance require the approval of committees responsible for evaluating all individual research programs in light of the company's numerous other opportunities and obligations.

In the end, some of these partnerships work extremely well for both parties. Others turn out less successfully, and many such arrangements are terminated each year, sometimes simply because the larger company has opted for a different strategic direction for its R&D focus. Either way, however, potential and actual corporate partners are highly influential constituents for virtually any biotech management team seeking to assemble and manage all of the resources necessary to bring a new medicine from the lab bench to the commercial market. In doing so, entrepreneurs will need not only to communicate their interests and goals effectively but also to understand, as well as possible, the forces driving decisions within the large company.

Physicians, Patients, and Payers

For most new biotech medicines that are launched around the world, the person receiving the benefit—the patient—neither makes the decision to buy it nor pays for it (at least directly). It will be prescribed by a physician and then will usually be paid for by an insurance company or national health service. The ultimate payer, whether it is private or public, will influence how the product is used by setting limits on what it will pay for, either with respect to the drug itself or for the overall treatment of patients with that condition. In Chapter 8, we will argue that decisions made by payers will have a very significant effect on the future of the industry.

More immediately, the needs and interests of physicians, patients, and payers will often influence the company's development decisions, especially in the design and conduct of clinical trials (typically the most expensive thing a company ever does). With health budgets increasingly strained, companies have become more aware of the need to demonstrate that their products are not only medically valuable but also cost-effective. According to a 2017 survey of 213 C-level executives and 87 healthcare investors conducted by Lazard Partners, pricing pressure was identified as the top strategic challenge, with the majority of respondents expecting fundamental changes in the traditional pricing models of the industry.[31] Yet demonstrating such evidence, especially in an evolving market in which cost-effectiveness goals may be moving

---

[31] Specifically, value-based pricing (where the payment is determined by the actual effectiveness of the drug on patients) and, more broadly, risk-sharing models between the payers and the manufacturers are expected to replace the current pricing model (which is primarily a fixed price per unit). See Lazard, "Lazard Releases Global Healthcare Leaders Study," news release, May 15, 2017, https://www.lazard.com/media/450 175/lazard-releases-global-healthcare-leaders-study-51517.pdf; and Tracy Station, "What Keeps 88% of Biopharma Executives Up at Night? One Guess, and It's Pricing," *Fierce Pharma*, May 18, 2017, https://www. fiercepharma.com/pharma/what-keeps-88-biopharma-execs-up-at-night-one-guess-and-it-s-pricing.

targets, may require larger—or even different—clinical trials than those that might lead to the most rapid regulatory approval.

Meanwhile, during the clinical development process (and often well before), companies will seek to encourage key opinion leaders (KOLs), who are typically the heads of departments or other well-known medical researchers in major university medical centers, to become knowledgeable about the product, to participate in the clinical trials, and to communicate the results through scientific presentations and peer-reviewed articles. Those leading physicians, in return, will often want to have a hand in designing the study, and, as with investors, not all KOLs in a particular field will always agree with one another on the best path forward. Patient advocacy groups can also influence the trials, sometimes by helping identify patients and clinical centers and sometimes by pushing for access to the product on a "compassionate use" basis even before regulatory approval, all of which requires the entrepreneur's time and attention to align all of these constituencies as much as possible.

The FDA and Other Regulatory Bodies

Perhaps the most influential constituencies a biotech entrepreneur will contend with are the regulatory bodies whose approvals are needed for a new medicine to be commercialized, such as the FDA and the European Medicines Agency. These government bodies influence corporate decision-making directly through their regulations as well as by the specific guidance they give to companies during clinical testing. They also have what might be called an "atmospheric" effect, in which the thinking and actions of essentially all of the other major stakeholder groups are influenced not only by what these agencies do but also by what the stakeholders think they might be inclined to do in the future. This phenomenon adds another complicating factor for biotech entrepreneurs since not all of the various constituencies—shareholders, partners, KOLs, etc.—will necessarily agree on their predictions about what the regulatory agencies will do.

Conclusion

Given the complex interactions among the stakeholders described in this chapter, in particularly challenging times, managing a development-stage, publicly traded biotech company can feel like trying to play three-dimensional chess against multiple opponents outdoors in an intermittent windstorm. The answers to the biggest questions in small biotech companies either are un-knowable and unpredictable—such as "Will this product successfully treat cancer?"—or otherwise involve making multidisciplinary decisions under conditions of great uncertainty.[32]

Consider a cash-poor biotech company that is trying to decide whether it should test its most advanced product in a small, single-arm Phase II trial with fifty patients or whether it should choose instead to invest much more time and resources in sponsoring a larger one with hundreds of patients and mul-tiple treatment arms. This question involves input from virtually every area of the company's activities. It requires balancing the medical needs of patients, the nature of the commercial marketplace, the relative position of competing products that are already on the market or are still in development, a regula-tory strategy that has the potential to lead to FDA approval, the availability of capital in an unpredictable financing market, current and future investors' expectations about the timeline for obtaining meaningful clinical data, the views (and decision rights) of any current or potential corporate partners, the availability of medical centers willing and able to participate in the clinical trial, competition for the same patients from clinical trials of other products, the company's organizational capacity, the availability of CRO resources, and a host of other factors.

To make a sound decision, the finance experts need to understand the clinical issues, the clinicians must appreciate the roles of both the regula-tory agencies and the company's inevitably shrinking cash position, and so on, with the answers to the most crucial questions being unpredictable, espe-cially for small companies with limited cash resources: Will the company have enough funding to survive to the end of the clinical trial, and will the results be good? All of these issues are complex, fascinating, and important, but fi-nally deciding on a particular plan of action is not necessarily a task for the faint of heart.

Beyond the many uncertainties inherent in the business that were discussed earlier, exogenous factors—market crashes or bubbles, changes in regula-tory or patent policies, and so on—can dramatically change the state of play and even the rules of the game. Frantic periods during financing bubbles, or when long-awaited data appear, are matched by other times of extended quiet, when the financial markets cool off, the clinical trial data are not expected for

---

[32] While entrepreneurial ventures in all industries are characterized by high levels of uncertainty, the bi-otech industry is different from nearly all of the rest because its major uncertainties cannot be resolved by low-risk/cost experimentation. For instance, in many other industries, start-ups can engage early on with customers to better understand what product or service attributes the customers really value. By doing so, major uncertainties can be resolved before the entrepreneur commits to high-risk actions.

months (or sometimes years), and interest from potential corporate partners has inexplicably dried up. Yet at any moment, the quiet, steady progress can be interrupted by the unexpected.

The business of biotechnology is often fascinating and frustrating, with the future prospects as bright as a life-saving, blockbuster drug and as dim as a greater than 90% failure rate and the difficult financing environments that are sometimes called "nuclear winters." Biotech entrepreneurs clearly share many characteristics with pioneers in various other fields of endeavor, but the extremely high costs, long timelines, and complex regulations associated with drug development add degrees of difficulty that are unusual, if not unique. For the right individuals, biotech entrepreneurship is constantly fascinating, often exhilarating, and regularly frustrating but ultimately can be both personally and professionally rewarding. Our goal in this chapter has been to help potential entrepreneurs think about whether it might be a good fit for their interests, whether their interests are a good fit for the industry, and why communications skills are so crucial for success. In highlighting the diversity of the tasks and stakeholders that a biotech entrepreneur needs to engage with, we have sought to encourage entrepreneurs to understand each group well enough to be able to understand what motivates them. Rarely does the biotech entrepreneur have the upper hand in these discussions, so success comes most often to those who can make arguments that resonate with someone who thinks like a stakeholder—a big pharma executive or a government regulator, for example. No matter what education, work experience, or personality traits a biotech entrepreneur may bring to the job, nurturing the ability to understand the motivations and goals of the key stakeholders will increase the likelihood of success in an industry that needs all the help it can get.

7
The Biotech Industry Track Record

At the outset of the COVID-19 pandemic, scientists around the world set out to design an effective vaccine as rapidly as possible. No one knew what technological approach would be the most successful. The goal was to prompt vaccinated people's bodies to generate a strong immune response to an essential part of the virus called the "spike protein." Then, upon exposure to the virus, the immune response would neutralize or eliminate it before the infection could set in. The key question was how to get the spike protein in the body in a way that would lead to a rapid, safe, and effective immune response.[1]

No one was better positioned to answer that question than GlaxoSmithKline (GSK), the world's leader in vaccines. GSK had over $8 billion in vaccine sales in 2019[2] and had opened a dedicated global vaccines research center just a few years before.[3] In those new labs, GSK researchers had been working with a promising new vaccine technology called mRNA. With that approach, they could potentially deliver the genetic code for the spike protein directly to the patients' cells. To get the resources to work on an mRNA-based COVID vaccine, the researchers, following the standard approach common to many pharmaceutical companies, presented the idea to the series of GSK committees that needed to approve research programs. After the proposal finally found its way to senior management, it was rejected on the grounds that mRNA technology was not "ready for prime time."[4]

At the same time, two small biotech companies set out to design mRNA COVID vaccines. Later recounting that it took one of them an afternoon,[5]

---

[1] For an overview of the technologies, see Aria Bendix, "Pfizer's Coronavirus Vaccine Relies on a New, Unproven Technology," *Business Insider*, November 9, 2020, https://www.businessinsider.com/leading-us-coronavirus-vaccines-how-they-work-compare-2020-10.

[2] GlaxoSmithKline, "GSK Delivers 2019 Sales of £33.8 billion +10% AER, +8% CER (Pro-forma +4% CER*)," press release, February 5, 2020, https://www.gsk.com/en-gb/media/press-releases/gsk-delivers-2019-sales-of-338-billion-plus10-aer-plus8-cer-pro-forma-plus4-cerstar/.

[3] GlaxoSmithKline, "GSK Opens New Global Vaccines R&D Center in Rockville, MD, USA," press release, December 13, 2016, https://www.gsk.com/en-gb/media/press-releases/gsk-opens-new-global-vaccines-rd-center-in-rockville-md-usa/.

[4] Andrew Dunn, "GlaxoSmithKline Stumbled with COVID-19 Shots," *Business Insider*, April 12, 2021, https://www.businessinsider.com/gsk-vaccine-research-covid-departures-uncertain-future-2021-4.

[5] Susie Neilson, "The Co-Founder of BioNTech Designed the Coronavirus Vaccine It Made with Pfizer in Just a Few Hours over a Single Day," *Business Insider*, December 12, 2020, https://www.businessinsider.com/pfizer-biontech-vaccine-designed-in-hours-one-weekend-2020-12.

and the other just two days,[6] the biotech companies pressed forward with the unproven technology. One of them, Moderna, developed the product on its own, while the other, BioNTech, partnered its vaccine with Pfizer for ongoing development.

Eventually, the Moderna and BioNTech/Pfizer vaccines were authorized for sale in many countries and became the leading COVID vaccines.[7] Analysts projected total 2021 sales of over $13 billion for the Moderna vaccine and over $19 billion for the BioNTech/Pfizer vaccine.[8] These biotech companies' willingness to bet on an unproven, potentially not-ready-for-prime-time technology propelled them well past GSK, which would no longer be the number one vaccine company.

This story illustrates many of the themes of this chapter. Biotech and pharmaceutical companies frequently have access to the same technologies and are working on the same diseases. But there are very significant differences in how decisions are made to allocate resources to the various research programs, leading to different levels of tolerance for risk. Those differences can help explain why the biotech industry has been able to create more high-priority, pathbreaking medicines despite having less experience and far fewer resources than the pharmaceutical industry.

Who Has Developed the Most Innovative New Medicines?

For most of the twentieth century, the pharmaceutical industry was recognized as a global leader in corporate research. Its consistent track record allowed managers of large pharmaceutical companies to devote an impressive 15–20% of gross revenues to R&D. During the years we analyze in this chapter, 1998–2016, pharmaceutical companies spent over $1.5 trillion on R&D, and there were five pharmaceutical companies ranking among the top fifteen companies in any industry in research spending in 2018.[9] This industry would

---

[6] Susie Neilson, Andrew Dunn, and Aria Bendix, "Moderna's Groundbreaking Coronavirus Vaccine Was Designed in Just 2 Days," *Business Insider*, December 19, 2020, https://www.businessinsider.com/moderna-designed-coronavirus-vaccine-in-2-days-2020-11.

[7] Kathy Katella, "Comparing the COVID-19 Vaccines: How Are They Different?" Yale Medicine, June 16, 2021, https://www.yalemedicine.org/news/covid-19-vaccine-comparison.

[8] Matt Egan, "Pfizer and Moderna Could Score $32 Billion in Covid-19 Vaccine Sales—in 2021 Alone," CNN, December 11, 2020, https://www.cnn.com/2020/12/11/business/pfizer-vaccine-covid-moderna-revenue/index.html.

[9] On the five pharmaceutical companies, see Barry Jaruzelski, Robert Chwalik, and Brad Goehle, "The Global Innovation 1000: What the Top Innovators Get Right," *Strategy + Business*, October 30, 2018, 7, accessed online, https://www.strategy-business.com/media/file/sb93-What-the-Top-Innovators-Get-Right.pdf.

seem unusually well positioned to fend off challenges from newcomers, especially in light of the formidable barriers to entry, including the complex array of FDA regulations. It was, therefore, not obvious that it would face the kind of entrepreneurial competition that emerged in computing and information technology. Nevertheless, a large and diverse array of biotech companies has grown up alongside the pharmaceutical industry. In this chapter, we evaluate how the newcomers have performed by benchmarking the R&D performance of the biotech industry against that of the pharmaceutical industry.

As we discussed in Chapter 1, Professor Gary Pisano's initial judgment, based on a similar comparison as of 2004, was that the biotech industry had been no more effective in research productivity than the pharmaceutical industry. He called it a "dead heat" at that point, and he observed that the biotech industry seemed to be veering toward devoting more resources to lower-risk, less innovative research.[10] Now that the biotech industry has had more time and capital to deploy in its R&D efforts, our evaluation of all new medicines approved over nearly two decades comes to a very different conclusion.

We have tracked the origins of the new medicines that were approved by the FDA from 1998 to 2016, and, in particular, we have measured the number of products from each industry that received "priority review" versus "standard review." Priority approvals are issued by the FDA for the subset of new drugs that not only treat serious or life-threatening conditions but also provide a significant improvement over existing treatments.[11] When the public thinks about medical breakthroughs—something that moves medicine forward in an important way—these priority review products are likely to be the discoveries it has in mind.

In pursuing the development of these high-priority new medicines, both the biotechnology and the pharmaceutical industries have employed the same basic tools of drug development: small-molecule chemistry, which has been in use for a century or more, and the newer biological tools, such as monoclonal antibodies and recombinant DNA, as we discussed in Chapter 2. While both industries have the same drug development goals and both use the same

---

[10] See Gary P. Pisano, *Science Business: The Promise, the Reality, and the Future of Biotech* (Brighton, MA: Harvard Business School Press, 2006), 120.

[11] The FDA describes the difference between priority and standard review as follows: "In 1992, under the Prescription Drug User Act (PDUFA), FDA agreed to specific goals for improving the drug review time and created a two-tiered system of review times—*Standard Review* and *Priority Review*. A Priority Review designation means FDA's goal is to take action on an application within 6 months (compared to 10 months under standard review)." Products eligible for priority review designation are those "that, if approved, would be significant improvements in the safety or effectiveness of the treatment, diagnosis, or prevention of serious conditions when compared to standard applications" (U.S. Food & Drug Administration, "Priority Review," last modified January 4, 2018, https://www.fda.gov/patients/fast-track-breakthrough-therapy-accelerated-approval-priority-review/priority-review).

tools to create potential breakthrough medicines, a key difference is that the decisions about which research programs will be provided with the necessary funding are made in very different ways. A fairly small number of pharmaceutical companies spend the lion's share of the industry's total R&D funding, and, within each company, one central committee typically controls all of the allocations. As a result, the research funding decisions for most of the R&D funds budgeted by the entire pharmaceutical industry are made by a fairly small number of committees. The degree to which so many of these decisions are concentrated in the hands of just a few committees contrasts sharply with the way biotech companies obtain funding from investors. In the biotech industry, thousands of independent companies compete for funding, which is provided by thousands of investors in the sector.

It turns out that economists and management scholars have devoted a considerable amount of theoretical attention over the past sixty years to the question of what approaches to R&D management and resource allocation are the most effective, especially in environments where there is a high degree of ambiguity. The creation of new treatments for serious medical conditions represents essentially a model case of performing research in ambiguous conditions. In this chapter, we will briefly outline what should happen in theory and then describe what actually happened in fact, which is—as the theory predicts—that the biotech industry developed significantly more priority products at considerably lower cost than the pharmaceutical industry. Following that analysis, we conclude the chapter by developing a framework that can help large pharmaceutical companies consider how they could put in place some practices and processes that may enhance their ability to discover novel medicines more efficiently.

The Impact of Ambiguity

Breakthrough medical products for unmet medical needs are almost inevitably developed in environments of high ambiguity. Decision makers not only face risk (that is, a low likelihood of success) but also, more importantly, lack of awareness about how reliable any estimates of potential success may be or what might be the attributes that lead to success.[12] That inability to know in advance what a successful outcome will "look like" is what makes the environment ambiguous. Hence, in contrast to an uncertain environment, where

---

[12] See Daniel Ellsberg, "Risk, Ambiguity, and the Savage Axioms," *Quarterly Journal of Economics* 75, no. 4 (November 1961): 643.

probability estimates carry significant value and can guide the decision-making process, in an ambiguous environment, such estimates can be highly misleading.

For example, before becoming a multi-billion-dollar blockbuster, first-in-class drug, Viagra was considered a low-priority drug at Pfizer, and the project was about to be terminated until researchers observed unexpected side effects suggesting that this type of drug could be effective for a completely different condition. Then, a subsequent drug, Cialis, which was in the same chemical class of drugs as Viagra, was developed for the same condition in a much less ambiguous environment because the link between that drug class and the medical condition had already been established. In that case, there was still uncertainty regarding the potential of Cialis itself, but the R&D teams had a much better understanding of the pathways for treating erectile dysfunction and therefore could make reasonable judgments about whether Cialis was moving in the right direction.

Economist Richard Nelson pointed out in 1961 that in environments dominated by ambiguity, initiating many parallel trials is essential for generating novel products.[13] Since then, a number of theoretical simulation models have shown that the parallel execution of nearly random, multiple trials outperforms complex optimization methods.[14] While this pursuit of multiple goals might seem costly at first sight, it outperforms a strategy that focuses early on one goal that then turns out to be the wrong one.[15]

Drawing on analogies from Darwinian evolution, management scholars have argued that the effectiveness of parallel trials in creating novel products relies on two key principles: generating a sufficient variety of projects at the outset, and then applying ruthless selection criteria ex post.[16] The variety of projects is necessary to ensure that the innovation funnel remains wide and is not prematurely narrowed before the unknown unknowns are resolved. Instead, by applying a parallel-trials approach, the decision makers maintain their option to select the most promising project once the key performance drivers are understood. For that reason, the parallel-trials approach is also referred to as selectionism. In the remainder of this chapter, we use the two terms interchangeably.

---

[13] See Richard R. Nelson, "Uncertainty, Learning, and the Economics of Parallel Research and Development Efforts," *Review of Economics and Statistics* 43, no. 4 (November 1961): 353.

[14] See, for example, Michael T. Pich, Christoph H. Loch, and Arnoud De Meyer, "On Uncertainty, Ambiguity, and Complexity in Project Management," *Management Science* 48, no. 88 (August 2002): 1008–23, and references therein.

[15] See Svenja C. Sommer and Christoph H. Loch, "Selectionism and Learning in Projects with Complexity and Unforeseeable Uncertainty," *Management Science* 50, no. 10 (October 2004): 1334.

[16] See Pich, Loch, and De Meyer, "On Uncertainty," 1015.

Research has also shown that the number of projects an organization decides to explore is directly affected by how its resource allocation process is structured. The seminal framework of Sah and Stiglitz[17] shows that environments with decentralized resource allocation processes result in wider project portfolios while centralized hierarchies lead to a narrower and less diverse funnel of opportunities. This theoretical model has recently been empirically validated in a large study of the mutual funds industry.[18]

Not only do organizations with central control over resource allocation tend to focus on fewer opportunities, but centralized decision-making environments are less likely than decentralized ones to terminate poorly performing projects.[19] In fact, it is often because companies with centralized decision-making processes are less capable of terminating bad projects that they initiate fewer projects in the first place.[20] Moreover, the inability of large centralized organizations to terminate exploratory projects effectively, even after unsatisfactory progress, can also lead them to devote resources only to projects that are more incremental in nature and appear to be less risky.[21] Therefore, the inability to launch a large array of projects at the outset, combined with a reluctance to terminate the failing ones efficiently, makes it difficult and expensive to create innovative products in ambiguous conditions.

A key reason for failing to cut off the funding for floundering programs and instead escalating the company's commitment to them by "throwing good money after bad" is the sunk-cost fallacy.[22] Managers falling victim to the sunk-cost fallacy allow the amount of resources already invested in a project to influence their opinions of the future value. Then, as the accumulated losses increase further, the project manager associates a termination decision with an even more detrimental negative balance, thus encouraging activities and analyses that will allow the termination decision to be further postponed.[23] It has been shown that large organizations are especially prone to the

[17] See Raaj Kumar Sah and Joseph E. Stiglitz, "The Architecture of Economic Systems: Hierarchies and Polyarchies," *American Economic Review* 76, no. 4 (September 1986): 716–27.

[18] See Felipe A. Csaszar, "Organizational Structure as a Determinant of Performance: Evidence from Mutual Funds," *Strategic Management Journal* 33, no. 6 (June 2012): 611–32.

[19] See, for example, M. Dewatripont and E. Maskin, "Credit and Efficiency in Centralized and Decentralized Economies," *Review of Economic Studies* 62, no. 4 (October 1995): 541–55.

[20] See Yingyi Qian and Chenggang Xu, "Innovation and Bureaucracy under Soft and Hard Budget Constraints," *Review of Economic Studies* 65, no. 1 (January 1998): 151–64.

[21] See Ramana Nanda and Matthew Rhodes-Kropf, "Innovation Policies," in *Advances in Strategic Management: Entrepreneurship, Innovation, and Platforms*, ed. Jeffrey Furman et al. (Somerville, MA: Emerald Publishing, 2017), Vol. 37, 60.

[22] Barry M. Staw, "Knee-Deep in the Big Muddy: A Study of Escalating Commitment to a Chosen Course of Action," *Organizational Behavior and Human Performance* 16, no. 1 (June 1976): 30.

[23] For a more detailed discussion about the sunk-cost fallacy, see Richard H. Thaler, *Misbehaving: The Making of Behavioral Economics* (New York: W. W. Norton, 2015), chap. 3; and Daniel Kahneman, *Thinking, Fast and Slow* (New York: Penguin Books, 2012), chap. 32.

effects of the sunk-cost fallacy, especially when the managers in charge of the project associate the success or failure of that project with their own career progression.[24]

In contrast to a large, centralized environment that can be prone to limiting the overall number of projects and can then be slow to stop the unsuccessful ones, a decentralized environment of multiple external investors maximizes the potential for following the two critical principles of (1) initiating many diverse projects and (2) stopping the ones that are not working out. Having many different decision makers—reflecting a range of risk tolerances, opinions, and judgments—who are responsible for allocating funding creates a favorable environment for trying many different things. It also minimizes the effects of the sunk-cost fallacy and the intra-organizational perspectives that make it difficult for large, centralized structures to make responsive termination decisions. Knowing that investors can quickly eliminate the underperforming investments[25] allows them to support a much wider and more diverse portfolio. In short, the theoretical arguments predict that a relatively small number of large organizations would tend to institute fewer parallel trials involving less innovative approaches (and would be less efficient in terminating unsuccessful ones) than an industry characterized by many independent investors making decisions about which companies and programs to support.

Taken together, these various streams of academic research suggest that the biotech industry's substantially more decentralized decision-making environment will generate more parallel searches and therefore a more diverse set of novel therapeutic product opportunities than the pharmaceutical industry and will be able to terminate the failing ones more efficiently. The result will be the more cost-effective creation of priority review products, which are the ones that involve the greatest degree of ambiguity. Not all drug development efforts are characterized by high levels of ambiguity, however. Second-generation products that build upon established development pathways might still entail risks, but those risks are much better understood. In such environments, it is critical to align the multiple functions of the organization (finance, human resources, marketing, operations) to a common goal (a well-defined

---

[24] See Eyal Biyalogorsky, William Boulding, and Richard Staelin, "Stuck in the Past: Why Managers Persist with New Product Failures," *Journal of Marketing* 70, no. 2 (April 2006): 108–21, and references therein.

[25] The importance of termination options for venture capital firms is discussed in William A. Sahlman, "The Structure and Governance of Venture-Capital Organizations," *Journal of Financial Economics* 27, no. 2 (October 1990): 473–521; and Steven N. Kaplan and Per Strömberg, "Characteristics, Contracts, and Actions: Evidence from Venture Capitalist Analyses," *Journal of Finance* 59, no. 5 (October 2004): 2177–210, and references therein.

target market, product specifications, a clear product concept) and ensure excellent communication and coordination among all the involved parties.[26] Thus, in those environments characterized more by uncertainty than ambiguity, the centralized structures of the pharmaceutical industry may be more effective than the collection of small independent biotech companies.

The Biotech Industry: A Textbook Case of a Decentralized Ecosystem of Parallel Searches

As we discussed in Chapter 4, the biotech industry consists of over five thousand companies, with the vast majority (around 86%) being private.[27] A 2010 report estimated that around 77% of all biotech companies had fewer than fifty employees.[28] This collection of small companies has created an extremely diversified portfolio of potential products that accounts for approximately 70% of the global drug development clinical pipeline: 3,706 of the 5,379 programs.[29]

Even among the publicly traded companies, resources are limited. The vast majority are unprofitable; around 25% of them operate with less than a year's worth of cash, while another 25% have less than two years' worth of cash.[30] Each company seeks to move toward FDA approval with whatever resources it can convince the investors to provide over time. A considerable variety of investors have contributed capital to these biotech ventures,[31] ranging from

[26] See Shona L. Brown and Kathleen M. Eisenhardt, "Product Development: Past Research, Present Findings, and Future Directions," *Academy of Management Review* 20, no. 2 (April 1995): 343–78. See also Lisa C. Troy, Tanawat Hirunyawipada, and Audesh K. Paswan, "Cross-Functional Integration and New Product Success: An Empirical Investigation of the Findings," *Journal of Marketing* 72, no. 6 (November 2008): 132–46.

[27] According to Ernst & Young's 2016 Biotechnology Industry report, in 2015 there were 2,772 companies in the United States and another 2,259 companies in Europe alone. See Ernst & Young, *Beyond Borders: Returning to Earth—Biotechnology Report 2016* (London: Ernst & Young, 2016), 18, 20.

[28] See Organisation for Economic Co-operation and Development, *Key Biotechnology Indicators*, December 2011, 1, accessed online, http://www.oecd.org/science/inno/49303992.pdf.

[29] See David Thomas and Chad Wessel, *Emerging Therapeutic Company Investment and Deal Trends* (Washington, DC: Biotechnology Industry Organization, 2016), 30, accessed online, http://go.bio.org/rs/490-EHZ-999/images/BIO_Emerging_Therapeutic_Company_Report_2006_2015_Final.pdf.

[30] We looked into Ernst & Young's reports during the time period of our data set, and we found that this percentage hovers between 20% and 30%. See Ernst & Young, *Beyond Borders: Matters of Evidence—Biotechnology Industry Report 2013* (London: Ernst & Young, 2013), 24; Ernst & Young, *Beyond Borders: Reaching New Heights—Biotechnology Industry Report 2015* (London: Ernst & Young, 2015), 16; and Ernst & Young, *Beyond Borders: Staying the Course—Biotechnology Industry Report 2017* (London: Ernst & Young, 2017), 33.

[31] By combining data from various sources, we identified nearly 5,000 firms, funds, and individual investors responsible for deciding which biotech companies get the resources to continue to develop their products: 561 venture capital/private equity firms, 245 mutual funds that specialize in the sector, over 3,000 generalist mutual funds, and 1,413 angel investors. The detailed list is available upon request.

venture capitalists and fund managers to business angels, potential patients, and the many individuals who "take a flyer" on a potentially hot biotech IPO. As of 2017, there were 561 venture capital and private equity firms, over three thousand mutual funds, and thousands of angel investors investing in biotech companies. These diverse, largely independent, investors have created an uncoordinated, decentralized decision-making environment for allocating capital across thousands of biotech companies. It is this combination of thousands of independent biotech companies being funded through the independent decisions of thousands of investors (which, together, we refer to as the biotech ecosystem) that has created a model for decentralized drug development based on parallel search.

Biotech investors expect many companies to fail.[32] These investors are the main drivers of the ruthless selection process, the second key principle of parallel search. Typically, they provide just enough capital for the firm to reach the next milestone.[33] The investors' goal is to quickly cut their losses in case of unsatisfactory results. Such capital-efficient termination decisions are common in the biotech industry, where the investors are constantly focused on the results of the next "killer experiment," which, in industry jargon, is one specifically designed to challenge a project by aiming to disprove its hypothesis rather than confirm it.[34] According to one venture capitalist:

> Innovation generally means a bit whacky and that then implies that it's got quite a lot of risk in it and the chances are it won't work. This is where we came up with the notion of the killer experiment, where the default is to kill the thing, not to continue with them. We run serial killer experiments where if we can't kill them at the first experiment, and it still looks good then we'll arrange another set of trials and if we can't kill it there, we keep doing it.[35]

---

[32] It is important to note that these investors do not diversify so widely because they think that there are so many good opportunities, but precisely because they realize that they cannot distinguish between "winners" and "losers" at that stage. According to a recent report that examined the distribution of the outcomes of 7,976 venture capital deals across venture sectors, 68% were money losing, and only 32% had some kind of positive returns. See Bruce Booth, "Correlation's Fresh Look at Venture Capital Returns," *Forbes*, November 18, 2013, https://www.forbes.com/sites/brucebooth/2013/11/18/correlations-fresh-look-at-venture-capital-returns/#7af4a0e040e5.

[33] See Sean Nicholson, "Financing Research and Development," in *The Oxford Handbook of the Economics of the Biopharmaceutical Industry*, ed. Patricia M. Danzon and Sean Nicholson (New York: Oxford University Press, 2012), 63.

[34] Venture capital firms sometimes invest in single-product companies and sometimes in multiproduct companies. In either case, given the companies' need for capital in order to continue their research programs, venture capitalists have a strong influence in the companies' product-funding decisions.

[35] Biotech and Money, "Interview: Killer Experiments and Keys to a Successful Deal," *Drugs & Dealers Magazine*, January 2015, 23, accessed online, http://cdn2.hubspot.net/hub/378634/file-2301474614-pdf/DDJan2015PDF.pdf.

Resource Allocation Processes in the Pharmaceutical Industry

For our benchmarking analysis, we defined the pharmaceutical industry as the eighty-five companies founded before 1976 that created at least one FDA-approved product between 1998 and 2016.[36] Almost all of the largest twenty companies were founded more than seventy-five years before the beginning of the biotech industry, and together they were responsible for about 80% of all pharmaceutical research spending, which was over twice as much as the biotech industry's total spending during the same time period. In sharp contrast to the decentralized decisions made by the thousands of investors providing funds to the biotech industry, resource allocation decisions in the pharmaceutical industry are made by a much smaller number of decision makers, not only within individual companies but also across the entire industry.

To better understand the resource allocation processes in the pharmaceutical industry, it is important to note that there is usually a corporate-level review committee that determines which projects receive funding, even in early-stage discovery research. This has been demonstrated in several case studies examining the resource allocation processes of the leading pharmaceutical firms,[37] and it was also highlighted to us in our interviews with senior pharmaceutical executives.

For instance, an executive from one of the largest companies noted, "These companies look huge from the outside, but they are actually very small inside, with only a handful of decision-makers." Another big pharma executive further highlighted the highly centralized nature of the selection process: "There is a quarterly review of all research projects by an eight-member committee, but, for any one project, there were really only two people who mattered—the head of that therapeutic area and the head of research."

Accordingly, the decisions that are necessary to initiate or terminate research and development projects within the pharmaceutical industry are typically made by a single corporate-wide committee within each individual company. Moreover, since there are only eighty-five companies, those

---

[36] As the result of mergers and acquisitions during the time of this study, there were only fifty-six companies remaining at the end of the period.

[37] See, for example, Robert S. Huckman and Eli Strick, "GlaxoSmithKline: Reorganizing Drug Discovery (A)," Harvard Business School Case 605-074 (May 2005), https://www.hbs.edu/faculty/Pages/item.aspx?num=32341; and James Mittra, "Impact of the Life Sciences on Organisation and Management of R&D in Large Pharmaceutical Firms," *International Journal of Biotechnology* 10, no. 5 (November 2008): 416–40. For additional information on the decision-making processes of several pharmaceutical companies, see Catrina M. Jones, "Managing Pharmaceutical Research and Development Portfolios: An Empirical Inquiry into Managerial Decision Making in the Context of a Merger" (EDB diss., Georgia State University, 2016), 61.

decisions are concentrated in the hands of far fewer decision makers in the pharmaceutical industry as a whole than in the biotechnology industry. Because of mergers and acquisitions among the largest companies, as of the end of our study, project approval and termination decisions relating to 71% of the pharmaceutical industry's research spending were being made by the research committees of the twelve largest companies.

Although some pharmaceutical companies have experimented with creating smaller, more decentralized R&D units—such as the Centers of Excellence in Drug Discovery (CEDD), and later the Discovery Performance Units (DPU), of GlaxoSmithKline—they have typically reverted to more centralized decision-making.[38] Even if some pharmaceutical companies adopted a more decentralized internal approach for a period covered by our study, the pharmaceutical industry as a whole would still represent a considerably more centralized system for resource allocation decisions than the large collection of investors and companies in the biotech ecosystem.

Comparing Biotech and Pharmaceutical Productivity

Based on the theoretical literature, we would predict that the structure of the biotech industry, with many independent investors allocating funding to a large number of parallel projects, would be more successful in generating the most pathbreaking products. The pharmaceutical industry, in filtering potential technologies and products based on their expertise in existing scientific knowledge ("fighting the last war," according to one executive), would be less well equipped to create novel products in ambiguous environments, especially where failures are very likely. For pathbreaking products, reliance on historical precedents will not necessarily be effective because the novel products' attributes of success are unpredictable. By contrast, a decentralized environment is not confined by such core rigidities[39] because of the independent nature of its units.

At the same time, the centralized nature of R&D within a large pharmaceutical company would suggest that it will be better able to coordinate

[38] See John Carroll, "GSK Déjà Vu: Time for a Top-Down Switch as New CEO Tries to Conquer Old R&D Demons," *Endpoints News*, August 7, 2017, https://endpts.com/gsk-deja-vu-time-for-a-top-down-switch-as-new-ceo-tries-to-conquer-its-old-rd-demons/.

[39] Dorothy Leonard-Barton, "Core Capabilities and Core Rigidities: A Paradox in Managing New Product Development," in "Strategy Process: Managing Corporate Self-Renewal," eds. Dan Schendel and Derek Channon, special issue, *Strategic Management Journal* 13, S1 (Summer 1992): 111–25.

information and resources between its discovery division and the rest of the organization (for example, the divisions that deal with the commercial and regulatory requirements). Such coordination is vital for the creation of second-generation products that have to be designed from the outset to compete in a competitive market, and therefore the pharmaceutical industry would be better equipped to originate such products.

A common limitation of many R&D performance studies is the lack of a commonly accepted metric of how to measure the innovativeness of a new product. We sought to overcome this challenge by looking at an objective criterion that applies equally to all the companies: the status of first-ever FDA approval of new molecular entities (NME).[40] In particular, we examine whether a product receives a "standard" approval, which means that it is sufficiently safe and effective for approval but not necessarily better than existing drugs, or whether it has been designated a "priority" product by the FDA, which means that it addresses an unmet medical need. An example of a standard review product is Celexa, approved for depression in 1998. There were already several approved drugs in the same SSRI class, including Prozac (approved ten years earlier). Meanwhile, cancer drug Yervoy received priority review approval. It was the first in the class of checkpoint inhibitors and the first drug to show a survival benefit for patients with advanced melanoma.

Companies seeking to develop priority review drugs encounter high levels of ambiguity. These are first-in-class medicines taking advantage of unprecedented and not always fully understood pharmacological pathways. For these products, there may be no reliable way to know in advance whether any one approach is more likely to succeed than any other. Meanwhile, standard review products typically need to compete with existing treatments that have already been shown to be safe and effective in addressing the same medical condition. Especially for me-too and second-generation products, the coordination of expertise among several functions is critical, even early in the research process. For example, scientists need to work with the legal and patent departments to fashion a new molecule that evades the competitor's patents without diminishing the product's effects, and commercial expertise is necessary to assess what product characteristics will be necessary to achieve

---

[40] According to the FDA, "Certain drugs are classified as new molecular entities ('NMEs') for purposes of FDA review. Many of these products contain active moieties that have not been approved by FDA previously, either as a single ingredient drug or as part of a combination product; these products frequently provide important new therapies for patients" (U.S. Food & Drug Administration, "New Drugs at FDA: CDER's New Molecular Entities and New Therapeutic Biological Products," last modified January 10, 2020, https://www.fda.gov/drugs/development-approval-process-drugs/new-drugs-fda-cders-new-molecular-entities-and-new-therapeutic-biological-products).

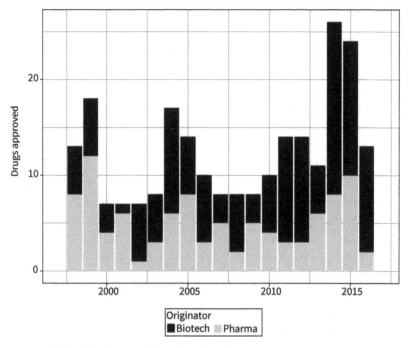

Figure 7.1 Original developer of priority review drugs
Source: www.fda.gov, EvaluatePharma, annual reports, and patent filings.

adequate market share. This process relies on aligning the capabilities of the complex array of scientific, medical, legal, manufacturing, marketing, and sales capabilities of the large companies.

Findings from Two Decades of Breakthrough Innovation

From 1998 through 2016, the FDA approved a total of 524 new molecular entities for sale as therapeutic products: 237 were selected for priority review and 287 for standard review. Of the priority review products (see Figure 7.1), 138 originated from biotech companies and 99 from pharmaceutical companies. That is, the biotech companies originated nearly 40% more FDA-approved priority products. Moreover, they did so while spending considerably less. We estimate that the biotech industry spent $2.97 billion per priority review product versus the pharmaceutical industry's $4.35 billion. (For a description of the methodology we used to arrive at these figures, see the "Postscript" section at the end of the chapter.)

The disruptive effect on the commercial market from these new entrants is clear: seven of the ten products with the highest revenues in 2019 originated in the biotech industry.[41] Two biotech companies—Gilead and Amgen—now have revenues ranking them among the dozen largest drug development companies in the world.[42]

The role of ambiguity in this discovery process and the importance of the parallel search approach are reflected in the wide range of different companies responsible for these innovative products. The 237 successful priority products were originated by a total of 146 different companies: 98 biotech companies and 48 pharma companies. Such success is not easily repeated, as even the most productive companies managed to originate only a handful of priority products in nearly twenty years.

While ownership of a product might change hands prior to FDA approval, as we discussed in Chapter 5, it is not the case that most innovative products created by biotech companies are ultimately commercialized by pharmaceutical companies. Of the products created by biotech companies in our analysis, only 28% were ultimately commercialized by pharmaceutical companies, with the rest remaining in the biotech industry. As a result, and as confirmed by our conversations with executives, the pharmaceutical companies have not abandoned innovative drug research with the expectation of just taking advantage of the products developed by the biotechnology industry.

With respect to standard review products, the pharmaceutical industry originated 180, and biotech companies originated 107 (see Figure 7.2). Out of those 107 biotech-originated products, 68 were commercialized by biotech companies, and 39 by pharmaceutical companies. The biotech ecosystem that generated more products in ambiguous conditions did not generate as many products aiming to enter a competitive marketplace with a complicated patent landscape,[43] an environment for which a highly coordinated collection of resources can be a significant asset.

As observed by a senior pharmaceutical executive who has also served as the CEO of a biotech company, standard review products provide an opportunity for the pharmaceutical companies to put their funding, size, expertise,

[41] See, for example, Kyle Blankenship, "The Top 20 Drugs by Global Sales in 2019," *FiercePharma*, July 27, 2020, https://www.fiercepharma.com/special-report/top-20-drugs-by-global-sales-2019.

[42] For the largest pharmaceutical companies, ranked by sales, see William Looney, "Pharm Exec's Top 50 Companies 2016," *Pharm Exec*, July 26, 2016, https://www.pharmexec.com/view/2016-pharm-exec-50.

[43] See Hangyu Zhao and Zongru Guo, "Medicinal Chemistry Strategies in Follow-on Drug Discovery," *Drug Discovery Today* 14, nos. 9/10 (May 2009): 516–22.

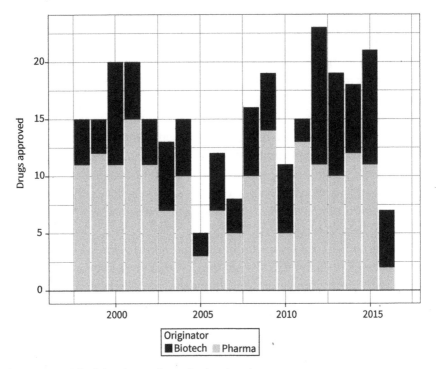

Figure 7.2 Original developer of standard review drugs
Source: www.fda.gov, EvaluatePharma, annual reports, and patent filings.

and other internal resources to good use by creating a multifunctional plan leading to a second-generation product and then systematically measuring progress against that plan.

The Importance of Running Many, Many Parallel Searches

The 237 successful priority products are important to consider, but it appears that the makeup of the rest of the R&D iceberg may be even more important. There are literally thousands of biotech companies whose new ideas did not pan out. But in the end, despite the failures, or more precisely, *because* of those numerous failures, biotech's many parallel searches generated more novel, priority review medicines at a lower cost than the pharmaceutical industry. To get a better understanding of this parallel search process, we used Thomson Reuters's Cortellis database to count the number of all R&D

programs[44] initiated by each industry from 1999 to 2016. While the pharmaceutical industry initiated 8,377 programs, the biotech industry initiated 39,619 programs—almost five times as many.

In an effort to enhance the effectiveness of their innovation research efforts, some of the large pharmaceutical companies have tried to restructure their central research functions to create small R&D units that have been designed to mimic the biotech industry,[45] and they have sought to access a wide range of opportunities through open innovation initiatives and collaborations with a plethora of universities and biotech companies.[46] Yet despite accessing a wider range of promising ideas, most pharmaceutical companies have not substantially widened their innovation funnel. An analysis of the collective R&D pipeline of the top fifteen (by 2016 sales) pharmaceutical companies shows that the total number of new programs (including both in-house and in-licensed programs) to which resources were allocated has remained approximately the same: around seven hundred new programs per year (including programs in clinical trials as well as those in advanced stages of pre-clinical research) for the last seventeen years (2000–2016).[47]

Management Approaches to Fostering More Breakthrough Innovation

As predicted by the theoretical literature discussed earlier in the chapter, our study of breakthrough medicines demonstrates that, in environments characterized by ambiguity, a widely diversified parallel search approach is the most effective strategy for originating radical innovation products. That is an excellent description of the biotech industry as a whole. The question that follows from this analysis is what steps the pharmaceutical industry can potentially take to become more effective in developing the most innovative medicines. Almost twenty years ago, Pisano recommended that pharmaceutical companies maintain a "vertical integration" approach to managing their most innovative research initiatives because his early review of the biotech industry's

---

[44] A program here is defined as a drug that enters a clinical trial for a specific indication.

[45] See Emma Dorey, "GlaxoSmithKline Presents a Biotech Façade," *Nature Biotechnology* 19, no. 4 (April 2001): 294–95.

[46] All major pharmaceutical companies have a website emphasizing open innovation and collaboration with external partners.

[47] See Figure C2 in Panos Markou, Stylianous Kavadias, and Nektarios Oraiopoulos, "Rival Signals and Project Selection: Insights from the Drug Development Process," working paper, *SSRN* (June 29, 2020): 62, https://ssrn.com/abstract=3225056.

track record reached a very different conclusion than we have about where the most innovative and important new medicines are created.[48] Our more recent data regarding research productivity suggest that the pharmaceutical industry should consider ways to take greater advantage of decentralization and selectionism. In particular, pharmaceutical companies may benefit from widening their innovation funnel by experimenting with products that would have been considered, under existing prioritization methods, too risky at that stage.[49] To the extent to which they may want to adopt a broader and more diversified parallel search approach, the first step is a good understanding of the key implementation barriers that are typically present in large organizations.[50] According to the academic literature and the pharmaceutical executives we interviewed, there are several such barriers in large organizations.

First, a well-established literature in innovation management has shown that even when established firms have access to an abundance of technological opportunities, they often demonstrate technological myopia[51] and select the ones that are similar to their existing product lines and competencies.[52] Novel ideas and projects that do not fit with the existing portfolio are viewed as "outlandish."[53] As one executive commented on past attempts of pharmaceutical companies to decentralize the funding and selection process: "At [name of company] they tried decentralization for a few years, but abandoned it when the lab heads were doing projects that management thought were too risky." Another executive also described how the existing pipeline might give rise to biases that affect future selection: "When you are sitting in a big pharmaceutical company, what you are hearing is a lot of people's own experiences of

[48] Pisano, *Science Business*, 177.

[49] Several recent studies have shown that project selection panels are likely to filter out the most novel ideas, even when those panels consist of highly qualified experts. See, for example, Kevin J. Boudreau et al., "Looking across and Looking beyond the Knowledge Frontier: Intellectual Distance, Novelty, and Resource Allocation in Science," *Management Science* 62, no. 10 (October 2016): 2765–83; Paola Criscuolo et al., "Evaluating Novelty: The Role of Panels in the Selection of R&D Projects," *Academy of Management Journal* 60, no. 2 (April 2017): 433–60; Laura J. Kornish and Karl T. Urlich, "The Importance of the Raw Idea in Innovation: Testing the Sow's Ear Hypothesis," *Journal of Marketing Research* 51, no. 1 (February 2014): 14–26; and Kyle Siler, Kirby Lee, and Lisa Bero, "Measuring the Effectiveness of Scientific Gatekeeping," *Proceedings of the National Academy of Sciences* 112, no. 2 (January 2015): 360–65.

[50] See, for example, Deborah Dougherty and Cynthia Hardy, "Sustained Product Innovation in Large, Mature Organizations: Overcoming Innovation-to-Organization Problems," *Academy of Management Journal* 39, no. 5 (October 1996): 1120–53; and Rita Gunther McGrath, "Exploratory Learning, Innovative Capacity, and Managerial Oversight," *Academy of Management Journal* 44, no. 1 (February 2001): 118–13.

[51] See John Wyman, "SMR Forum: Technological Myopia—the Need to Think Strategically about Technology," *Sloan Management Review* 26, no. 4 (Summer 1985): 59–64.

[52] See Michael L. Tushman and Philip Anderson, "Technological Discontinuities and Organizational Environments," *Administrative Science Quarterly* 31, no. 3 (September 1986): 439–65; and Clayton M. Christensen, *The Innovator's Dilemma: When New Technologies Cause Great Firms to Fail* (Boston: Harvard Business Review Press, 2016).

[53] Roberto Verganti, "The Myths That Prevent Change," *Harvard Business Review*, May 16, 2012, https://hbr.org/2012/05/the-myths-that-prevent-change.

what went wrong in the past. In biotech, you're making it up as you go along, and you're not limited by those biases."

Second, recent literature has shown that having access to a wider pool of opportunities does not necessarily lead to more novel ideas being selected. This happens because a wider pool might overwhelm the limited number of decision makers. As another senior executive noted:

> The real problem in big pharma is that the noise to signal ratio is so huge. The question is how to get the right information at the right time to the small number of people who can make decisions. Having too many resources can lead to bad decisions. You become so swamped by the filtration system you need for so many opportunities that you end up increasing the noise, making it harder to find the signal.

As a result, and rather paradoxically, access to more resources and opportunities might make an organization more risk-averse in its selection process.[54] Senior executives charged with the task of assessing a plethora of opportunities might become more inclined to prefer options closer to the status quo.[55] Project managers, who often compete for resources, are bound to become risk-averse and, in turn, only recommend incremental projects to senior management, as those are the ones they believe will be more likely to get approved.[56] Hence, under a highly centralized resource allocation structure, even if promising ideas exist somewhere in the organization, the most novel ideas are likely to be filtered out.

Third, in our discussions with pharmaceutical executives, they expressed concerns that a wider funnel of opportunities will result in excessive resources being wasted in projects that eventually fail. This happens because it is so hard to terminate projects once they get started. This tendency is succinctly summarized by one executive who noted that "Once everybody, finally, is on board for a project, it takes an act of God to stop it even if things are not going well." The natural question is: Why is it so hard to run "killer experiments" under centralized organizational structures? A different pharmaceutical executive shed some light on this question: "It's hard to kill things because people don't want to admit to failure. It lets other business units point to your failure when

---

[54] See Barry Schwartz, *The Paradox of Choice: Why More Is Less* (New York: HarperCollins, 2004); and Roberto Verganti, *Overcrowded: Designing Meaningful Products in a World Awash with Ideas* (Cambridge, MA: MIT Press, 2017).

[55] See Donald A. Redelmeier and Eldar Shafir, "Medical Decision Making in Situations That Offer Multiple Alternatives," *JAMA* 273, no. 4 (January 1995): 302.

[56] See Amit Seru, "Firm Boundaries Matter: Evidence from Conglomerates and R&D Activity," *Journal of Financial Economics* 111, no. 2 (February 2014): 384.

they are competing for resources." Moreover, terminating underperforming projects in large companies is often hard because of the reward and promotion structures, which are focused on interim outcome-based milestones and encourage "weather the storm" behaviors rather than admissions of failure. As another executive noted, "The company [has] a 'champion' culture where people were rewarded for pushing projects forward."

Lastly, it is interesting to note the crucial interdependency between the two stages of project selection and efficient termination decisions. Given that senior management is worried that all programs will be expensive and take up a lot of resources, it often decides to control costs by starting fewer projects. As a result, the innovation funnel becomes narrower and less diversified, leading to a risk-averse selection process that does not encourage the creation of radical innovation products.

A Framework for Widening the Innovation Funnel

Building on our benchmarking analysis, our in-depth discussions with highly experienced pharmaceutical executives, and the emerging academic literature on the management of innovation that highlights the role of incentives for experimentation,[57] we have developed a framework that could help pharmaceutical companies to implement a more diversified parallel search effort. This framework suggests specific practices that they could consider in their efforts to overcome the implementation barriers discussed in the previous section. We refer to this innovation strategy as SMART (Selectionism MAkes Research Transformative) to emphasize that it is driven by the knowledge creation process rather than the firm's existing assets and capabilities. To highlight the contrast, we also present the corresponding suggested practices for companies that decide to focus their R&D resources in less ambiguous environments. We refer to this strategy as FAST (Focused and Accelerated with Standard Tests) to emphasize that the R&D process is driven by and focused on the firm's well-established capabilities (see Figure 7.3 and Table 7.1).

First, for the project selection stage, our SMART framework offers managers an alternative performance metric that they could consider adopting to assess the progress of those radical projects for which standardized evaluation methods perform poorly or are unavailable: the diversity and volume of opportunities under consideration. According to recent literature, one way to

---

[57] See, for example, Gustavo Manso, "Creating Incentives for Innovation," *California Management Review* 60, no. 1 (November 2017): 18–32.

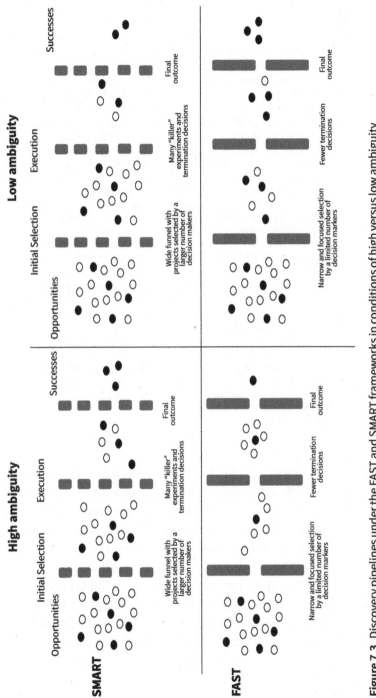

Figure 7.3 Discovery pipelines under the FAST and SMART frameworks in conditions of high versus low ambiguity

Note: In high-ambiguity environments, a FAST approach narrows the innovation funnel too early (at the initial selection stage), based on results that are not yet predictive of the ultimate success. By contrast, a SMART approach allows for many more opportunities to enter the funnel but terminates them quickly with the right "killer experiments." In low-ambiguity environments, early results are more predictive of the ultimate success, so the more focused selection of FAST (at the initial selection stage) allows companies to be more able to construct candidates that are likely to succeed.

Table 7.1 The SMART and FAST frameworks

	Breakthrough products SMART	Incremental products FAST
Organizational structure	Resource allocation decisions made by a large number of decision makers with diverse perspectives and independent decision rights	Resource allocation decisions made by a small number of decision makers that set the strategic priorities and objectives of the selection process
Selection of projects	• Reward volume and diversity of project opportunities • Allow people to explore risky ideas without requiring consensus from the rest of the organization • Acknowledge the presence of ambiguity and do not give too much weight to information that has not been demonstrated to be a good predictor of success • Refrain from narrowing the funnel too quickly	• Prioritize projects based on quantitative criteria or "fit" with company's strategic position and capabilities • Develop contingency plans to account for the effects of uncertainty • Cross-functional coordination and information sharing to improve fit with strategy
Managing the project portfolio	• Plan for failure: experiments should be designed to challenge the working hypothesis, not to confirm it • Assess projects through a diverse panel of people who can scrutinize information from different angles • Include external experts to avoid vested interests • Reward people who fail fast and efficiently by running the right "killer" experiments	• Plan for success: projects are expected to deliver their objectives and meet the criteria • Reward good performance and successful outcomes • Adjust the project portfolio to maximize synergies with current products

achieve this is by increasing the number and the diversity of people involved in the selection decision.[58] Moreover, the breadth of the innovation funnel is affected by the decision rule (that is, the requirements to get a project approved) that a company applies during the selection process.[59] For example, if unanimous approval is required for a project to receive funding, only a very limited number of the opportunities will receive approval; it takes only one executive to consider it "too risky" for the project to be terminated. Conversely, a decision rule that requires just one "yes" from a senior executive for initiating a project maximizes the diversity of the innovation funnel. This is precisely the approach used by Amazon, for example, to ensure that if an executive is excited about an idea, he or she is given the opportunity to pursue it.[60]

---

[58] See Criscuolo et al., "Evaluating Novelty," 433–60.
[59] See note 9.
[60] See "Ten Rules of Innovating at Amazon," Stories, Keller Center at Princeton University, last modified January 18, 2018, https://kellercenter.princeton.edu/stories/ten-rules-innovating-amazon.

Second, as discussed earlier, the successful implementation of a parallel search approach relies on fast and efficient project termination decisions. Lovallo and Kahneman's influential article about the "Delusions of Success" suggests that an effective way for dealing with the sunk-cost fallacy is for companies to establish diverse project review panels (including external experts) that can scrutinize the project from different perspectives.[61] The key is to actually design and implement "killer experiments" and to seek evidence that disproves a project's hypotheses, rather than eliminate it simply because it does not fit with the company's existing product lines. In their efforts to become more efficient at project terminations, pharmaceutical companies may also draw valuable insights from the literature on flexibility in product development[62] and agility in project management[63] that highlight how fast and constant feedback not only speeds up the project execution but also, more importantly allows the project team to adjust the project as more information becomes available.

Lastly, it is important to recall that a parallel search process is likely to lead to a large number of failures, especially in the process of new drug development. Incentive systems that reward and promote managers on the basis of their recent success records are not compatible with an environment that fosters a parallel search strategy. Instead, because fast and frequent failures are the precursors of novel products in an environment of ambiguity, organizations could benefit by creating incentive mechanisms that encourage them. This is exactly the rationale behind X's (formerly Google X's) approach to award a bonus (and even a holiday trip) to teams that decide to kill their own projects.[64]

The mindset required for testing the principles of the SMART framework would be a major departure from the status quo of many large companies whose decision-making structure is closer to the FAST framework. Therefore, the first step for large companies aiming to develop radically innovative products is to evaluate whether the SMART framework is compatible with their organizational culture and risk tolerance as well as the career expectations of their managers.

[61] See Dan Lovallo and Daniel Kahneman, "Delusions of Success: How Optimism Undermines Executives' Decisions," *Harvard Business Review* 81, no. 7 (July 2003): 56–63.

[62] See Marco Iansiti and Alan MacCormack, "Developing Products on Internet Time," *Harvard Business Review* 75, no. 5 (September 1997): 108–17.

[63] See Jim Highsmith, *Agile Project Management: Creating Innovative Products*, 2nd ed. (Upper Saddle River, NJ: Addison-Wesley, 2009); and Jake Knapp, John Zeratsky, and Braden Kowitz, *Sprint: How to Solve Big Problems and Test New Ideas in Just Five Days* (New York: Simon & Schuster, 2016).

[64] See Jessica Leber, "How Google's Moonshot X Division Helps Its Employees Embrace Failure," *Fast Company*, April 14, 2016, https://www.fastcompany.com/3058866/how-googles-moonshot-x-division-helps-its-employees-embrace-failure.

Companies seeking to commercialize radical products, but for which the organizational culture would not be a good fit with the SMART framework, can consider other ways to participate in a more diversified parallel search process than is found in their in-house R&D programs. For example, numerous pharmaceutical companies have adopted open innovation initiatives to accomplish this goal.[65] By outsourcing some or all research targeting radical innovation by investing in venture capital firms or acquiring options on early-stage products being developed by small companies, the pharmaceutical companies can let external funders decide what risks to take and when to kill things off.[66]

Conclusion

In this chapter we compared the track records of the biotech and pharmaceutical industries. Our key finding is that the biotech industry has been responsible for the creation of more pathbreaking medicines than the pharmaceutical industry, and it has done so at a much lower overall cost. Our results offer the first major reassessment of the productivity of the biotech industry in nearly two decades, and they provide the first large-scale empirical validation for the mathematical models that have shown that in highly ambiguous environments, the parallel execution of nearly random initiatives outperforms complex optimization methods. This strategy is the most effective because it can maximize the opportunities for success by taking advantage of new knowledge as it develops stochastically and often serendipitously, by creating an environment that will encourage trying many things, and by ruthlessly abandoning the ones that do not work out.

Corporate R&D teams in large, experienced companies may be the best in the world at managing product development pathways, but in highly ambiguous environments, they are often inclined to make decisions based on today's

---

[65] More information about the industry report can be found at John Carroll, "Big Pharma Is Using Its Venture Cash to Outsource Early R&D to Biotech," *Fierce Biotech*, July 31, 2014, https://www.fiercebiot ech.com/venture-capital/big-pharma-using-its-venture-cash-to-outsource-early-r-d-to-biotech. For a recent example of early-stage outsourcing, see Angus Liu, "Takeda, Schrödinger Form Multiprogram Drug Discovery Pact," *Fierce Biotech*, July 20, 2017, https://www.fiercebiotech.com/cro/takeda-schrodinger-form-multiprogram-drug-discovery-pact.

[66] Recent academic literature has highlighted the importance of leveraging the power of outsiders and the need to adapt the mode of collaboration to the strategy of the company. See Gary P. Pisano and Roberto Verganti, "Which Kind of Collaboration Is Right for You?," *Harvard Business Review* 86, no. 12 (December 2008): 78–86, in which they discuss different types of collaboration modes and the importance of selecting the right one. They also emphasize cases where innovation solutions required transferring critical decision rights to the partner companies.

knowledge, much of which will change in unpredictable ways. In that environment, R&D management needs to avoid falling into the trap of narrowing the funnel too quickly based on unreliable estimates and assumptions. The risk of such a derisking process is that the next blockbuster product is forgone because it did not fit today's strategic criteria.

Our framework lays out some basic guidelines on how large organizations can overcome those traps and widen their innovation funnels, but much more research is needed in this area. Decentralized environments have, by definition, a large number of independent decision makers. What should be a sufficiently large number for a large organization? How can the current hierarchical corporate structures accommodate such delegation of decision-making power? The extant academic literature offers little guidance on these topics, as it has been primarily focused on single projects rather than a portfolio of parallel searches. The transition to managing multiple projects raises other new topics of interest. First, what is the role of senior management and leadership in a decentralized organization where key selection and termination decisions are made by a large number of people? Second, how should the teams running those parallel projects be incentivized so that they make the best decisions for the organization rather than just their own team? These are challenging questions, but we believe that they give rise to several exciting prospects for future research into the innovation process that can have important real-world applications.

Postscript: How We Did the Benchmarking Analysis

With the assistance of Broad Brook Research, we built a novel data set that traces the company that originally created each one of the 524 priority review and standard review products that were approved by the FDA between 1998 and 2016.[67] Our data set integrates information from a number of different

---

[67] We are very grateful to Michael Breidenbach of Broad Brook Research for his outstanding efforts in creating this data set. Beginning in 2009, the FDA began awarding "priority review vouchers" to incentivize companies to develop certain types of products. As described in a Government Accountability Office analysis, the FDA will issue a priority review voucher to a company that obtains approval for a new product to treat a tropical disease or a rare pediatric disease. The transferable voucher can be used (by that company or any company to which it transfers the voucher) to obtain a priority review from the FDA for a subsequent product that has not been designated by the FDA as a priority product. See U.S. Government Accountability Office, *Drug Development: FDA's Priority Review Voucher Programs*, January 2020, 2, accessed online, https://www.gao.gov/assets/710/704207.pdf. Because we have used priority review status as an objective standard signifying that a product addresses an unmet medical need, we have not considered products approved on a priority basis solely because of the use of a voucher to be priority review products for the purposes of our analysis. For the same reason, we have excluded from the data set fourteen therapeutic proteins that were not assessed for priority status by the FDA under the same standards that applied to the remaining products. These fourteen were approved prior to the June 2003 transfer of therapeutic proteins from the FDA's Center of Biologics Evaluation and Research (CBER) to the Center for Drug Evaluation

sources: the FDA's website, commercial databases that track the R&D portfolio of pharmaceutical and biotech companies (for example, EvaluatePharma, Thomson Reuters's Cortellis), annual reports and 10-K fillings submitted to the U.S. Securities and Exchange Commission, and patent fillings. To the extent that some of those products were encompassed by other published studies (for example, reports published by the Tufts Center for the Study Drug Development and annual industry reports by Ernst & Young) or unpublished data to which we were able to obtain access, we cross-checked our identification of the originator of the product with those studies.[68]

Data about the R&D expenses of the pharmaceutical and public biotech companies were obtained from their annual reports and 10-K fillings, while data about the public and private funding raised by the biotech industry were collected from Ernst & Young's annual reports and Standard and Poor's Capital IQ database. The latter was also used to estimate the number of biotech companies, the number of investors, and so on.

We estimated the aggregate R&D spending for the years 1998 and 2015 for the pharmaceutical industry by totaling the R&D expenditures of the eighty-five companies. For the biotechnology industry, we combined the total R&D expenditures of all publicly traded companies with 100% of the amounts of venture capital invested in the biotech industry for the same years. During these years, the pharmaceutical industry spent a total of $1,635.52 billion on R&D, and the biotechnology industry spent $626.74 billion, both in 2014 dollars.

To estimate the cost per priority review product for each industry and consistent with earlier studies,[69] we made the following adjustments: (i)

and Research (CDER). See U.S. Food & Drug Administration, "Transfer of Therapeutic Products to the Center for Drug Evaluation and Research (CDER)," last modified February 2, 2018, https://www.fda.gov/about-fda/center-biologics-evaluation-and-research-cber/transfer-therapeutic-products-center-drug-evaluation-and-research-cder. In an independent analysis of these products, Kneller classified all fourteen as priority products. Ten of these products were created by biotech companies and three by pharmaceutical companies, with one resulting from a collaboration between a biotech and a pharmaceutical company. See Kneller, "Importance of New Companies," 868 and Supplementary information S2 table. If we had included these fourteen products in the data set as priority review products, the number of priority products created by biotech companies would be nearly 50% more than the number of priority review products originating in the pharmaceutical industry.

[68] See, for example, Robert Kneller, "The Importance of New Companies for Drug Discovery: Origins of a Decade of New Drugs," *Nature Reviews Drug Discovery* 9, no. 11 (November 2010): 867–82. Because the ownership of technologies and products can change over time, we have followed the following allocation rules. If the product is first created in an academic institution, we consider the originator to be the company that in-licensed it from the university. This is not meant to diminish the essential role played by academic research in the creation of important new medicines, but to allow us to compare the choices made by biotech and pharmaceutical companies. In cases where Company A creates a product and then licenses it to Company B (or is acquired by B), we credit A as the originator and B as the commercializing company. If A licenses an enabling technology to B (but not a product), and B creates a product using that technology, B is identified as the originator.

[69] See, for example, Pisano, *Science Business*, chap. 6.

subtracted from the pharmaceutical companies' total expenditure the 18% of R&D that industry executives have said is dedicated to the development of products that are not new molecules;[70] (ii) assigned to priority review products the percentage of total spending represented by the fraction of priority review products over total products; and (iii) approximated the allocable costs of products that moved from biotech to pharmaceutical companies (through licensing agreements or acquisitions), or vice versa, by assigning 50% of the costs to the originator and 50% to the commercializing company. After making these adjustments, we estimate that the pharmaceutical companies spent $4.35 billion and the biotech industry spent $2.97 billion for each priority review product approved by the FDA.[71] We employ this methodology, as others have done in the management literature,[72] to identify major differences in overall costs rather than to attempt to calculate the actual costs per product, which would require access to the confidential internal records of thousands of companies.[73]

These data were complemented by our discussions with executives. First, we interviewed seven senior executives from the drug development industry, who collectively have been employed by seven pharmaceutical companies (all currently in the twelve largest companies), four life sciences venture capital firms, and twelve biotech companies (see Table 7.2). During those interviews we asked them to reflect on our assumptions and estimates, explain the resource allocation process in the organization they worked for, and reflect on if and how a parallel search approach would be implementable in those organizations. Second, we studied the literature on the decision-making processes of pharmaceutical organizations, including consulting

---

[70] See Congressional Budget Office, *Research and Development in the Pharmaceutical Industry*, Pub. No. 2589 (Washington, DC: CBO, 2006), 21n7, https://www.cbo.gov/sites/default/files/109th-congress-2005-2006/reports/10-02-drugr-d.pdf.

[71] Given that biotech companies developed 138 priority review products and 107 standard review products, we allocated (138/237 = 56%) of the total R&D expenses of the biotech industry to priority review products. Similarly, for the pharma companies, we allocated (99/280 = 35%) of the total R&D expenses that were spent on NMEs. Consequently, the biotech industry spent $353 billion to originate 118.5 priority review products, while the pharmaceutical industry spent $402.4 billion to create 92.5 priority review products. As such, the cost per product is $2.97 billion for biotech and $4.35 billion for pharma. Lastly, note that if we were to use the same percentage for both the pharma and biotech companies, the allocated cost would be lower for the biotech companies and higher for the pharmaceutical companies. Thus, our category-specific ratio for the cost allocations is a conservative estimate regarding the cost efficiency of the biotech industry compared with the pharma industry.

[72] See Pisano, *Science Business*.

[73] As discussed in Chapter 3, the most extensive studies on the average cost for developing a new drug come from the Tufts Center for the Study of Drug Development, which are based on confidential survey data by a range of biopharmaceutical companies. Their model is based on a Monte Carlo simulation that encapsulates the average cost per phase for preclinical and clinical studies, the average transition probabilities per phase, development timelines, etc. Given the fundamental differences in the approaches used in their study versus ours, the figures are not comparable.

Table 7.2 Interviewee's role and company information

Respondent	Role	Industry experience	Firm age	Turnover
R1	Head of business development (BD)	Over 15 years	Over 100 years	ca. $50b
R2	Vice President BD	Over 30 years	Over 100 years	ca. $50b
R3	Senior Director (SD) R&D	Over 30 years	Over 100 years	ca. $30b
R4	SD BD, Biotech CEO	Over 15 years	Over 100 years	ca. $20b
R5	VC partner, Biotech CEO	Over 25 years	N/A	N/A
R6	SD BD, Biotech BD	Over 10 years	Over 100 years	ca. $50b
R7	Global head BD	Over 40 years	Over 100 years	ca. $50b

company reports and several case studies describing how resource allocation decisions were made.[74]

[74] See Nils Behnke and Norbert Hueltenschmidt, *Changing Pharma's Innovation DNA* (Boston: Bain, 2010), https://media.bain.com/Images/BAIN_BRIEF_Changing_pharmas_innovation_DNA.pdf; and James Mittra, "Impact of the Life Sciences on Organisation and Management of R&D in Large Pharmaceutical Firms," *International Journal of Biotechnology* 10, no. 5 (November 2008): 416–40.

8

The Future of the Biotech Industry

When an industry has been disrupted by new technologies and a flood of entrepreneurial challengers, predicting the future often revolves around the question of when the industry will settle down, consolidate around a few winners, and take on mature characteristics.[1] In the case of biotech and pharma, that prognostic approach would suggest that the large pharmaceutical companies would either adapt or die; the best of the biotech companies would join the successful adapters as the mainstays of a new status quo; and there would no longer be thousands of little drug development companies. But our analysis of the biotech industry's track record, as discussed in Chapter 7, suggests that the drug development business will resist this common cyclical pattern.

In many business sectors, consolidation around early winners offers competitively valuable economies of scale.[2] In contrast, our research shows that the development of important new medicines is considerably more efficient in the hands of large numbers of independent entities than when it is done internally by large companies, even when those companies have the advantages of greater scale, scope, experience, and resources. As long as drug development failure rates remain as high as they have been for decades—and the discoveries of the last half-century have not substantially moved that needle—we believe that large companies (even ones that entered the biotechnology arena as small companies) will be unwilling or unable to take enough product development shots on goal due to the organizational structures and career development incentives described in Chapter 7. Moreover, the impressive growth of the contract research industry means that the biotech companies that can raise enough cash to move their products forward will continue to be able to access all of the drug development resources they need.

If maturation and consolidation are unlikely prospects, at least for the next few decades, what will determine the future of the biotechnology industry?

---

[1] See Steven Klepper and Kenneth L. Simons, "Innovation and Industry Shakeouts," *Business and Economic History* 25, no. 1 (Fall 1996): 81–89.

[2] For an example from the banking industry, see Diana Hancock, David B. Humphrey, and James A. Wilcox, "Cost Reductions in Electronic Payments: The Roles of Consolidation, Economies of Scale, and Technical Change," *Journal of Banking & Finance* 23, nos. 2–4 (February 1999): 391–421.

Early in the book, we described biotechnology companies as "the offspring of a marriage of convenience between science and money." Peering into the future involves asking first about the science and, in particular, whether biomedical breakthroughs are likely to continue to arise. If so, then the key question is whether investors will keep pouring billions of dollars into the industry in the belief that those discoveries will continue to be profitable enough to justify the high-risk investments.

As to whether important scientific breakthroughs remain on the horizon, it would seem to go without saying that there is a great deal left to be learned about the nature of disease and the human body. The nearly immobile failure rates in drug development are just one sign that the amount we know today is likely to be a fraction of what we will be able to learn in the future. Moreover, the biomedical tools on which most of today's novel products are based—monoclonal antibodies and recombinant DNA—were invented in the 1970s, nearly fifty years ago. More recent technologies, from gene therapy to RNA interference (RNAi) to many others, remain in their relative infancy, and the many tools of artificial intelligence have hardly made a dent in our pharmacopeia. On the scientific side, there will be many more breakthrough opportunities. There could be many potential blockbusters on the investment side as well, but the future in that arena will be determined less by the promise of the science than by the resolution of much-debated questions of public policy. Why government policymakers will likely determine the future of what would seem, on its face, to be a quintessentially business-focused question of risk and return is the subject of most of this chapter. Interestingly, the most visible place where government enters the financial equation—the need for the approval of the FDA and similar regulatory bodies in other countries—will have far less impact than the central question of who will pay for new medicines and how much will they pay.[3]

The Investment Case

In 2010, there was a controversial proposal to reduce cancer research funding in the United Kingdom by over £100 million a year. The chief executive of

---

[3] Our goal in this chapter is to discuss the fundamental trade-offs that need to be considered in making those decisions rather than describing the numerous technical details and complexities of reimbursement policies. For a detailed discussion on the latter, see Robert E. Wanerman and Susan Garfield, "Biotechnology Product Coverage, Coding, and Reimbursement Strategies," in *Biotechnology Entrepreneurship: Leading, Managing and Commercializing Innovative Technologies*, ed. Craig Shimasaki, 2nd ed. (London: Academic Press, 2020), 499–512.

Cancer Research UK wrote an article for *The Times* in support of maintaining the funding. He did not begin his essay with the hopes and fears of desperate cancer patients and their families or even with the promise of Nobel prize-winning discoveries. Instead, he wrote, "Scientific research makes a vital contribution to the economy." To sharpen the financial point, the article was titled "Years of Crucial Investment Could Fall by the Wayside."[4]

The financial returns on scientific investment can appear so dramatic that the health benefits from biomedical research may pale in comparison. For example, in a report prepared for the State of New Jersey, leading economists from Rutgers University projected over $70 *billion* of direct and indirect returns from a comparatively modest $720 *million* government investment in one particular new medical technology.[5] Such eye-catching figures certainly make a persuasive case for public investment in the life sciences. The prospect of medical breakthroughs for patients with devastating illnesses, combined with these kinds of massive near-term and longer-term economic benefits, make for a compelling argument.

For the last several decades, these types of analyses have led governments in many countries to invest heavily in biomedical research. This massive investment in basic research provided the scientific foundation not only for the biotechnology industry's building-block technologies—such as recombinant DNA and monoclonal antibody technologies—but also for the discovery of many of the disease targets toward which those technologies have been directed. For instance, NIH-supported research helped create the first cancer drug targeted at a family of molecules called kinases. That discovery opened the way for an entirely new class of drugs leading to dozens of new medicines for cancer and other diseases.[6]

The public policy choices along this pathway have been fairly straightforward: Government funding creates new discoveries, often in the laboratories of thriving university biomedical science departments. Some of those discoveries lead to spin-off companies that attract private capital, thus creating jobs and economic growth. Some of those companies create new medicines, which help patients live longer, thus adding to economic growth. Some of those medicines are very profitable, which creates more jobs, more economic

---

[4] Harpal Kumar, "Years of Crucial Investment Could Fall by the Wayside," *Times* (London), September 8, 2010, 16. He also notes that the "U.S. and China are increasing investment in science, calculating that it will help to generate growth" (Kumar, "Years of Crucial Investment," 16).

[5] See Joseph J. Seneca and Will Irving, *Updated Economic Benefits of the New Jersey Stem Cell Capital Projects and Research Bond Acts* (New Brunswick, NJ: Rutgers, 2007), ii, v, accessed online, https://bloustein.rutgers.edu/wp-content/uploads/2015/03/stemcelloct07.pdf.

[6] See "Our Society," National Institutes of Health (website), last modified May 1, 2018, https://www.nih.gov/about-nih/what-we-do/impact-nih-research/our-society.

growth, and more taxes flowing back to the governments, which allows for continued investment in biomedical research, thus continuing the cycle.

At the same time, as with any economic analysis, making sense of the output figures requires an understanding of the underlying assumptions behind the calculations. To that end, it is worth taking a closer look at the report prepared for the State of New Jersey, which aimed to estimate the economic benefits of the planned tax-supported funding program for promoting research into the use of stem cell technology for the treatment of a number of life-threatening conditions. Both the original (written in 2005) and the revised (written in 2007) reports were prepared for the governor by Joseph Seneca and Will Irving of the Edward J. Bloustein School of Planning and Public Policy at Rutgers, the State University of New Jersey.

Through the application of what the authors termed "conservative assumptions throughout the report," Seneca and Irving forecasted financially compelling returns generated by the government's investment in this new technology.[7] The key element required to achieve the predicted economic growth was attracting nongovernmental, for-profit investment in the stem cell initiative. Of the projected $1.4 billion of new economic activity, over $1 billion was attributable to this "private leveraging."[8] The report assumed that for every $1.00 in public spending, venture capitalists (or other for-profit investors) would invest $1.50.[9] State money would prime the pump, but 60% of the funding would flow from private sources. If New Jersey were to adopt the initiative, the financial returns to the state would go well beyond the R&D jobs created, equipment purchased, taxes paid, and other direct benefits of the research initiative. The report identified many billions more in "broader benefits" from savings in future healthcare costs resulting from better treatments for Parkinson's disease, Alzheimer's disease, spinal cord injury, and the like ($11.3 billion in overall medical savings); "work time and economic productivity savings" ($813 million); and the "value of premature deaths avoided" ($60.7 billion).[10]

The report also sounded a cautionary note about the risks of not funding this particular area of research. In the 2005 report, the authors warned that the health of the state's existing biotechnology industry could be threatened if this stem cell initiative were not to take place because many states and

---

[7] Joseph J. Seneca and Will Irving, *The Economic Benefits of the New Jersey Stem Cell Research Initiative* (New Brunswick, NJ: Rutgers, 2005), iv, accessed online, https://pdfs.semanticscholar.org/5fdf/1098479d3 26f963d83caee3f30f94bb4b214.pdf.

[8] Seneca and Irving, *Economic Benefits*, i–ii.

[9] See Seneca and Irving, *Economic Benefits*, 37.

[10] Seneca and Irving, *Economic Benefits*, iii.

countries are competing for the "biotechnology industry, with its high value added and knowledge-based workforce."[11] If the "locus and focus of U.S. stem cell work, highly competitive to start, . . . centers on other locations," private capital will shift to where the most exciting research is taking place, and, the report argued, up to 20% of New Jersey's existing biotechnology jobs could move to the places where "the intellectual and commercial synergies associated with proximity to others in the industry attract the new start-ups."[12] Without the program, then, thousands of existing jobs were projected to be lost, along with the private capital that seeks the best new discoveries wherever they take place.[13] Because the biotech workforce is so highly educated, biotech (and pharma) employees are compensated at a level that is more than twice the average annual wage of private sector workers.[14] The proposed stem cell program, therefore, represented both an investment in the potential for new medicines and a technological "land grab" by which New Jersey could increase its national share of especially well-paying biomedical jobs.

In this competitive environment, New Jersey was not alone in seeking financial and technological benefits from investing in stem cell research. These types of stem cell economic analyses were being performed in a number of other states at roughly the same time, and they typically shared the same broad assumptions as the New Jersey study. A report issued by the California Stem Cell Research and Cures Initiative was prepared by a pair of researchers, one of whom was the Chief of Health Services Research at Stanford School of Medicine. The detailed analysis in support of California's multibillion dollar bond issuance for stem cell research identified "total state revenues and health care cost savings of between $6.4 billion and $12.6 billion . . . generating a 120% to 23% return on the [state's] investment made in the research."[15]

[11] Seneca and Irving, *Economic Benefits*, 32.

[12] Seneca and Irving, *Economic Benefits*, 36.

[13] For a summary of the "bidding war between states" to attract stem cell research, see Office of the New York State Comptroller, *The Economic Impact of the Biotechnology and Pharmaceutical Industries in New York* (New York, 2005), 2, accessed online, https://web.osc.state.ny.us/osdc/biotechreport.pdf. For example, "Voters in California recently approved a referendum to spend $3 billion in state funding over ten years to create the California Institute for Regenerative Medicine, eclipsing efforts by other states (including New York) to attract elements of the biotech industry and stem cell research" (NY State Comptroller, *Economic Impact*, 2).

[14] According to a Battelle/Bio report, based on the Bureau of Labor Statistics data, the "Drugs and Pharmaceuticals" sector had an average annual wage of $106,576 as of 2012, compared with a "Total Private Sector" wage of $49,130. See Battelle and Biotechnology Industry Organization, *Battelle/BIO State Bioscience Jobs, Investments and Innovation 2014* (Washington, DC: Biotechnology Industry Organization, 2014), 9, accessed online, https://www.bio.org/sites/default/files/legacy/bioorg/docs/files/Battelle-BIO-2014-Industry.pdf.

[15] Laurence Baker and Bruce Deal, *Economic Impact Analysis: Proposition 71 California Stem Cell Research and Cures Initiative* (Boston: Analysis Group, 2004), 2, accessed online, https://www.analysisgroup.com/uploadedfiles/content/news_and_events/news/proposition_71_report.pdf. For a discussion of some alternate scenarios, not all of which yield such promising economic outcomes, see Michael T.

A subsequent 2019 report that quantified the economic impact of the California Institute for Regenerative Medicine (CIRM) estimated that CIRM activities generated $15.4 billion of additional gross output (sales revenue) across the U.S. economy ($10.7 billion in California and $4.7 billion in the rest of the country) and have resulted in 82,365 additional jobs.[16]

A key common foundation in the analyses of both the New Jersey and the California reports is that the lion's share of these financial returns to the state was expected to be derived from two sources: *private investment* exceeding the amount of the state's funding and many billions of dollars in *reduced healthcare costs* as a result of the new stem cell treatments. These cost savings were based on the assumption that new therapeutic products emerging from the initiative would immediately lower, not raise, the cost of treating the patients. To make the cost savings analysis work, the new products must reduce the overall cost of treating a patient by at least 10% and reduce mortality by about 7%.

Attracting sufficient levels of private capital is the other essential component for generating the economic multiplier effect that increases the financial return from the state's investment in the near term. The reports do not say why they believe that private capital will follow the public dollars, although economists have observed that, in the past, increases in public spending on medical research has generated a corresponding growth in private R&D funding. Even if we assume that those studies are correct,[17] we need to consider whether these historical data are likely to predict the responses of future biotech investors. If not, these programs will have little chance of generating any of the projected billions in economic growth and other benefits that justify the government's continued investment in basic biomedical research.

Longaker, Laurence C. Baker, and Henry T. Greely, "Proposition 71 and CIRM—Assessing the Return on Investment," *Nature Biotechnology* 25, no. 5 (May 2007): 513–21.

[16] See Dan Wei and Adam Rose, *Economic Impacts of the California Institute for Regenerative Medicine (CIRM)* (Los Angeles: University of Southern California, 2019), ES-1, ES-2, accessed online, https://www.cirm.ca.gov/sites/default/files/CIRM_Economic%20Impact%20Report_10_3_19.pdf.

[17] See, for example, Andrew A. Toole, "Does Public Scientific Research Complement Private Investment in Research and Development in the Pharmaceutical Industry?," *Journal of Law and Economics* 50, no. 1 (February 2007): 81–104; and the studies summarized in the annotated bibliography contained in the HERG Report: Health Economics Research Group (HERG) Brunel University, Office of Health Economics (OHE), and RAND Europe, *Medical Research: What's It Worth? Estimating the Economic Benefits from Medical Research in the UK* (London: UK Evaluation Forum, 2008), 56–59, accessed online, https://mrc.ukri.org/publications/browse/medical-research-whats-it-worth/. It is certainly the case that during the 1960s, 1970s, and 1980s, the period examined during the Toole study, NIH funds and industry funding increased in tandem. Whether there is a direct causative effect is less clear, especially since Toole's data suggest an unlikely (in the opinion of practitioners) "U-shaped" response in which industry responds positively to increases in public spending in the first and second years after the public increase, and in the seventh and eighth years, but not in years three through six.

In summary, the economic case for governments to continue to fund basic biomedical research is based on the expectation that private investment will boost that funding by 150% or more. The key question for the future is: Will private sources of capital continue to invest in the biotechnology industry? As we write this chapter in the midst of the COVID-19 global health crisis, the answer is clearly yes. An effective vaccine or treatment will allow millions to go back to work, and national economies will be able to recover from the lockdowns instituted to slow the spread of the disease. For once, a complicated economic model is not necessary to show the link between scientific investment and financial return.[18] Once that crisis has run its course, the question of the continued attractiveness of biotechnology investment for both private investors and governments will return. That search for an answer will revolve around two core issues: Who will pay for new medicines, and, especially if the answer is that governments will be the primary payers, how much will they be willing and able to pay?

Critical Assumptions for Continued Investment in Biotechnology

As discussed in Chapter 4, venture capitalists are usually the first sources of money that fund new job growth at entrepreneurial companies. The general partners make the investment decisions and monitor the progress of their investments, but the funds themselves come primarily from large institutional investors. Government and corporate pension funds, as well as many university endowments, are constantly evaluating a very diverse array of possible investment opportunities. Their decisions to invest some of their funds (usually a small percentage of their total assets) in venture capital are purely financial ones. In return for committing to a (usually) ten-year venture capital fund that invests in a high-risk sector such as biotechnology, they expect to receive a considerably better return than they would get by investing in lower-risk industries.[19]

These institutional investors seek out the best-performing venture capital firms—in which performance is measured solely in terms of financial

---

[18] For an estimation of economic and job losses caused by COVID-19 as of May 2020—$2.1 trillion in lost income and 147 million jobs—see Manfred Lenzen et al., "Global Socio-Economic Losses and Environmental Gains from the Coronavirus Pandemic," *PLOS ONE* 15, no. 7 (July 2020): 1.

[19] A 2010 report showed that over 80% of worldwide investors in venture capital are in North America and Europe. See Preqin, *The Venture Capital Industry: A Preqin Special Report, October 2010* (London: Preqin, 2010), 9, accessed online, https://docs.preqin.com/reports/Preqin_Private_Equity_Venture_Report_Oct2010.pdf. Overall, over half of those investors are pension plans, endowment funds, and foundations.

returns—and they leave the investment decisions to the general partners. Pension and endowment funds are very sophisticated investors with billions to invest, and even if they plan to dedicate some of their assets to venture capital, they do not have to do so in the life sciences. In other words, just as biotech companies compete with one another to obtain venture funding, biotech venture capitalists compete with other types of venture capital firms for investments from pension and endowment funds. These funds can pick and choose their investments for what they believe will be the best possible financial returns.[20] They will therefore expect to be very well compensated for investing in something as risky, illiquid, and time consuming as biotechnology venture capital. As a result, venture capitalists in the life sciences must pick investment opportunities with prospects for high enough returns to compensate for the risks involved since many new drug prospects turn out to be failures, and those failures are often extremely expensive.

When everything goes well from the venture capitalist's investment to a profitable exit—by a sale to a pharmaceutical company or a high valuation on the public markets—a promising technology has the chance to be transformed into a new medicine; the economy has grown by the addition of new jobs; the pension and endowment funds have more money with which to fulfill their institutional missions of paying for people's retirements or funding academic programs; patients have obtained access to a medicine that might not otherwise have existed; and the public's health is improved, not only providing people with better health, but also allowing them to be more economically productive, thus further improving the economy. This is obviously a very attractive scenario.

The problem is that these happy endings occur so rarely. Only a minority (sometimes a remarkably small minority) of venture-backed biotech companies will be sold at a profit,[21] and therefore those profits need to be very substantial. To sustain venture investing in biotechnology, the venture capitalists, and especially their investors, need to be confident that the people making decisions at the pharmaceutical companies or investment funds that make those exits possible will continue to believe two things: first, that a product that beats these daunting odds will be approved for commercial sale by the regulatory agencies in the most important pharmaceutical markets; and second, that such an approval will create sales of the product at high enough

---

[20] See Paul A. Gompers and Josh Lerner, "What Drives Venture Capital Fundraising?," *Brookings Papers on Economic Activity. Microeconomics* 29, no. 1998 (July 1998): 149–204.

[21] Similarly, "Less than 1 percent of the Stanford disclosures have generated $1 million or more in cumulative royalties": National Research Council, *Managing University Intellectual Property in the Public Interest* (Washington, DC: National Academies Press, 2011), 23.

profit levels to provide an attractive return on the very substantial investment required to pay for this successful drug development effort *plus* pay for all of the unsuccessful ones. If these beliefs no longer seem reasonable, then the flow of money pouring into biomedical research will screech to a halt, and the associated economic growth will stop along with it.[22]

The Cost of New Drugs and Improved Healthcare

In fields outside the life sciences, technological advances often lead to ways to do things faster, better, *and cheaper*. In computer engineering, this phenomenon is often referred to as Moore's Law, based on Intel cofounder Gordon Moore's original statement that the "number of transistors incorporated in a chip will approximately double every 24 months."[23] This technological prediction has been expanded to encompass not just "doubl[ing] . . . transistor density" but simultaneously "increasing functionality and performance and decreasing costs."[24] Advances in healthcare seem mostly to tack in the other direction.

First, new medical products are likely to be much more expensive to create than new products in the information technology field. According to a 2008 report from RAND Europe, "the average out-of-pocket cost of a new [pharmaceutical] product approval is $400 million . . . whilst in other highly innovative sectors, such as IT and communications technologies, an investment of £4 million allows companies to bring new products to the market."[25] As discussed in Chapter 3, the cost of drug development has more than doubled in the last decade, with the most recent estimates putting the total at over $2.5 billion per product. Second, unlike new information technology products that deliver cheaper and better "solutions" to the user, biomedical discoveries have improved treatments, but they often come at significantly higher costs than the previous, less effective treatments. In many cases, then—and particularly in connection with products that extend life or create substantial improvements in people's quality of life—the novel products are often far more expensive than existing treatments.

---

[22] During IPO "windows," IPO investors will need to share these same beliefs about the possibility of profitable products, since they will either be investing in the expectation that the company will become profitable or (more likely) be acquired by a large pharmaceutical company.

[23] "Moore's Law and Intel Innovation," Intel (website), accessed September 17, 2020, https://www.intel.com/content/www/us/en/history/museum-gordon-moore-law.html?wapkw=moore%27s%20law.

[24] "Intel Celebrates 40 Years of Digital Revolution," Intel Newsroom, last modified November 15, 2011, https://newsroom.intel.com/news-releases/intel-celebrates-40-years-of-digital-revolution/#gs.gfpc6m.

[25] HERG, OHE, and RAND Europe, *Medical Research*, 83.

Prices for new medicines have reached levels that were likely unimaginable when the biotech industry was formed. A cancer patient's annual cost for combination chemotherapy might have been a few thousand dollars then, whereas a single new high-tech cancer product in the twenty-first century might be priced at $150,000 or more. A breakthrough drug for hepatitis launched by biotech leader Gilead Sciences in 2013 was so expensive, at over $80,000 for a course of therapy, that one report noted, "At a private meeting for payers and policy makers . . . the [Chief Medical Officer] of one of the big [payers] noted that if Gilead Sciences Inc.'s . . . HCV [hepatitis C] drug Sovaldi sofosbuvir were used in all eligible patients, it would add $1,000 to every health insurance premium in the U.S."[26] The reporter, industry veteran Roger Longman, commented that these cost increases would reduce future costs, however, because "the treatment prevented down-the-road complications of the disease."[27]

For our purposes, the crucial financial point is that biotech's breakthroughs may be lifesaving, but they have rarely been cost saving. At least at the outset, new and better medicines are likely to be more expensive—potentially much more expensive—than existing treatments. As a case in point, the recently approved drugs based on gene therapy make the Sovaldi example look relatively cheap. Priced at $2.1 million, Novartis's Zolgensma was granted the title of the most expensive drug in the world in 2019. A significant element of the pricing rationale was that Zolgensma can be a one-and-done, lifesaving cure for spinal muscular atrophy, a rare childhood disorder.[28]

Finding a novel therapy can therefore create a double challenge for the economy, whether the new treatment is for hepatitis or cancer or for neurological diseases such as Alzheimer's or spinal cord injury. Once it is proven effective, the new product may not necessarily replace existing inadequate treatments. As a result, brand new money is needed to pay for the brand-new treatments. Whether, as the New Jersey stem cell projections suggested, healthcare payers will allocate $3 billion to just one new (perhaps non-cost-saving) product over a seven-year period beginning eleven years after the investment is a difficult question.[29] Especially after the recession of 2008 (and

[26] Roger Longman, "Guest Commentary: Cost Steamroller," *BioCentury*, June 9, 2014, https://www.bio century.com/article/248976.

[27] Longman, "Cost Streamroller."

[28] For the most expensive drugs in the United States as of 2019, see Eric Sagonowsky, "The Top 10 Most-Expensive Meds in the U.S.—And They're Not the Usual Suspects," *Fierce Pharma*, June 13, 2019, https://www.fiercepharma.com/pharma/most-expensive-meds-u-s-topped-by-novartis-and-spark-gene-therapies.

[29] The 2007 New Jersey report notes that successful "biotechnology therapies are estimated to have generated $3 billion in sales per drug over the patent life." Seneca and Irving, *Updated Economic Benefits*, 17. The report then appears to assume that the $3 billion will be available from future healthcare budgets,

following), investors became increasingly worried about whether high prices are going to continue to be available for every new product. As Longman noted, "specialty products" such as Gilead's hepatitis drug "are now front and center in the mind of the drug decision-maker[s], whether insurer, physician or patient. And in particular, the people who pay for biotech's products are finally saying 'enough.' "[30]

Not surprisingly, how much insurance companies and governments are willing to pay for new treatments for serious diseases is a critical question that arises when biotech venture capitalists seek to raise new funds from their investors. And so, in considering the availability of the private capital that is necessary to allow the government investments in the life sciences to achieve the predicted financial gains to society, we need to bear in mind that venture capitalists invest not only in the dream of medical progress arising from new technologies but also in the belief that healthcare payers—largely insurance companies and governmental policymakers—will find the money to pay premium prices for the fruits of pioneering research.

Government Policy and the Market for New Medicines

The policymakers involved in deciding whether to pay for new medicines are not necessarily thinking about near-term economic growth. In the United States, for the most part, private payers such as insurance companies decide what their market—that is, their premium paying customers—will bear. They need to consider to what extent they can profitably sell health insurance policies at price levels high enough to pay for expensive new medicines. Customers willing to pay higher premiums will have more insurance coverage for expensive treatments, whereas those paying less for insurance may have to pay more (or all) of the costs out of their own pockets—if they can afford to do so. Some medicines may be affordable for only those who can pay for very expensive health insurance policies. As a result, rich and poor individuals will not necessarily have access to the same medical care.

In countries such as the United Kingdom and the numerous others that have national health systems, the situation is quite different. These systems provide the same healthcare for all, irrespective of income. As a result, they

presumably because it is a fraction of the $11.3 billion that the report assumes would otherwise have been spent on less effective treatments.

[30] Longman, "Cost Steamroller."

will pay for only products that they can afford to make available to any citizen who needs them. Because the national health budget is limited, some medicines are not approved for use in the national health system and may be available to only the patients who have opted to acquire private insurance or pay the costs of private clinics.

A national health system approach can have a significantly different effect on the nature of the commercial market for new medicines compared with the private insurance market, which is a key reason why the United States, with about 4% of the world's population, represented about 40% of the global pharmaceutical market in 2018.[31] In the United States, wealthy individuals (and others with insurance coverage with very high limits and low patient copays) represent a commercial market place for high-priced products. (It should be noted that many biopharmaceutical companies also have patient assistance programs that make expensive products available to low-income patients. It is likely that the costs of the patient assistance programs increase the price of the medicine for those who can afford to pay for it.)

From the perspective of the government policymakers considering how to parcel out a limited healthcare budget, the critical question is: what is the value (economic, philosophical, theological, or otherwise) of paying billions of dollars for what may be a modest-but-real benefit to patients with Parkinson's or Alzheimer's? That benefit is likely, at least in the new treatment's early years, to increase overall healthcare costs rather than decrease them, despite the assumptions in the New Jersey and California reports discussed earlier.[32] Those enthusiastic reports in support of government funding for stem cell research rely on the largely self-evident prediction that society will be willing to pay lower prices for better health. That is an easy question. The more likely and significantly harder question is how much *more* will we pay for the possibility of better and longer lives and, indirectly, for the R&D necessary to create the opportunity for those longer lives.[33]

[31] IQVIA Institute for Human Data Science, *The Global Use of Medicine in 2019 and Outlook to 2023: Forecasts and Areas to Watch* (Parsippany, NJ: IQVIA, 2019), 49, accessed online, https://www.iqvia.com/insights/the-iqvia-institute/reports/the-global-use-of-medicine-in-2019-and-outlook-to-2023.

[32] See Longaker, Baker, and Greely, "Proposition 71," 518, who state that "new stem cell therapies will not necessarily reduce spending; indeed, they may drive spending up.... If the costs of the new therapies were sufficiently high, there may not be any net healthcare savings. New therapies that significantly lengthen life could also elevate costs simply by increasing the amount of baseline health spending, though presumably also producing the benefit of longer life. Perhaps the most important possibility, however, is that stem cell research will make possible treatments that augment the set of therapies used for conditions, rather than replacing or obviating the need for other therapies. The history of healthcare innovation includes many instances where the development and improvement of treatments expanded the medical arsenal, rather than replaced older treatments. This increases health spending over time."

[33] See Kevin M. Murphy and Robert H. Topel, "The Economic Value of Medical Research," in *Measuring the Gains from Medical Research: An Economic Approach*, ed. Kevin M. Murphy and Robert H. Topel (Chicago: University of Chicago Press, 2003), 71, in which they state, "Expenditures on research that

We are not arguing in this chapter that either public or private payers must always decide in favor of higher spending on new high-technology drugs. Our point is simply that if they do not, then the much-anticipated "leverage" from private investment will likely disappear, and its multiplier effect on economic growth in the biomedical fields will quickly dissipate. That effect will also make it less attractive for governments to continue to fund basic research in the life sciences, and opportunities for innovative products will decline.

The Government Market for New Medicines

How much governments should be willing to pay for new medicines is a challenging question on how to allocate scarce healthcare resources fairly. In the United Kingdom, all medicines must meet fairly strict cost-effectiveness standards to qualify for payment from its National Health Service (NHS). As a result, some of the most promising new molecular medicines, including some that were discovered in the United Kingdom, have not been available for the many patients covered by the NHS. As just one example, we discussed in Chapter 2 the importance of monoclonal antibody technology to the biotechnology industry. Scientists at the government-funded Medical Research Council laboratories in Cambridge, England, won the Nobel Prize for discovering that technology, and it is likely that more entrepreneurial companies have been focused on monoclonal antibodies than on any other single technology emanating from biomedical research in the last fifty years.

On the one hand, this one technology has led to a great deal of the economic growth that the life sciences have stimulated over the past four decades, thus making antibodies one of the most vital drivers of all biotech-related economic growth. Yet monoclonal antibodies are often so expensive that the NHS and similar organizations in other countries have refused to pay for them. Whether these kinds of high-tech drugs are available to patients in the United Kingdom, for example, depends on a multi-step process that only partly relates to diagnosis or prognosis and does not appear to relate at all to economic growth. In other words, it appears that the basic research was funded out of one U.K. government pocket, but the resulting new medicine was not paid for out of the other. Hence the argument by Cancer Research UK at the beginning of the chapter: not only will U.K. lives be lost if cancer

---

prolong life without increasing expenditures are almost sure to be worthwhile . . . but expenditures that significantly increase costs could easily generate more expenditures than gains given the distortions in the care market (the prevalence of third-party payers and the annuitization of benefits)."

funding is cut, but also the economic value of medical research—something that the healthcare payers may not have considered—will be threatened.

Taxes set by Parliament fund the NHS, which turns to the National Institute for Health and Care Excellence (NICE) for advice on how to parcel out the inevitably limited resources available to pay for healthcare. In considering how this system works, it is important to bear in mind that a similar mechanism for approving the cost of new treatments is in place in a number of other countries and could someday also include the United States. As of 2020, a new product would be recommended by NICE for reimbursement if its cost is less than £20,000 for a "quality-adjusted life year" (a "QALY" in the trade), although for special cases this figure could go up to £30,000 per QALY or even higher.[34] The concept of the QALY is to combine "both quantity (length) of life and health-related quality of life into a single measure of health gain," thus providing a "common currency" for comparing different treatments for the full range of illnesses. That is, NICE will calculate how much it will cost for a new drug to extend a patient's life by one "quality year" (a fully active life being more valuable than the same year in a coma, for example), thus allowing the NHS to compare the value of a new cancer drug to a new treatment for heart disease. As a result of the QALY calculations, some of the potentially life-extending monoclonal medicines that were based on the discoveries of the Cambridge Nobelists do not qualify for NHS use because they cost too much.[35]

One side of the QALY equation is a fairly straightforward collection of costs: the new drug itself, any hospitalization required, medical personnel, whatever medications may be needed for dealing with likely side effects, and anything else that allows NICE to estimate the comprehensive cost of the treatment and its expected consequences. These numbers are then weighted according to the probabilities associated with the various events, as with a side effect that only requires medication a third of the time. There may be lively debates among participants in the relatively new field of pharmacoeconomics about how to do these allocations and calculations, but the question that looms over the entire analysis is: How should a government agency decide what a year of life is worth in an era of straitened resources for healthcare?

[34] According to NICE's website, "For treatments that extend life at the end of life, we can go as high as £50,000 per QALY. For treatments for very rare conditions NICE uses a different threshold range of between £100,000 and £300,000 per QALY" ("Our Charter," National Institute for Health and Care Excellence, accessed September 17, 2020, https://www.nice.org.uk/about/who-we-are/our-charter).

[35] One recent example is Roche's Polivy. See "NICE Turns Down Roche's Polivy for Lymphoma Combination," FDAnews, last modified March 5, 2020, https://www.fdanews.com/articles/196131-nice-turns-down-roches-polivy-for-lymphoma-combination.

The sensible and straightforward nature of these largely fact-based cost calculations should not obscure the really difficult moral and political judgments that are essential to the NHS's final decision about whether to pay for a new medication. NICE's recommendation, once implemented by the NHS, inevitably involves choosing whose "life years" will be extended and whose will not be. Will every citizen be allocated an equal amount of the national medical budget, or will the chronically ill get a bigger share? Will rich and poor receive equal opportunity for new medicines? Should both taxpayers and the unemployed be treated the same? Are some diseases less worthy of expensive treatments than others? Should saving lives trump treating debilitating but non-fatal conditions? These questions ask policymakers to reflect on the value of life and on how we deal with genuinely difficult choices when some lives may end up being lost in the interest of saving others.

These are challenging questions, and NICE released a document in 2005 titled "Social Value Judgements: Principles for the Development of NICE Guidelines" to address these issues.[36] It offers a valuable overview of the kinds of philosophical issues facing healthcare policymakers. The "Social Value Judgements" analysis acknowledges that NICE must make not only scientific value judgments but also social value judgments, which "relate to society rather than science."[37] It begins by listing four "widely accepted moral principles" that have been adopted because "they provide a single, accessible, and culturally neutral approach that encompasses most of the moral issues in healthcare."[38] The four principles are:

- respect for autonomy,
- non-maleficence,

---

[36] See National Institute for Health and Clinical Excellence, *Social Value Judgements: Principles for the Development of NICE Guidance*, 2nd ed. (London: NICE, 2005), https://www.nice.org.uk/Media/Default/about/what-we-do/research-and-development/Social-Value-Judgements-principles-for-the-development-of-NICE-guidance.docx. The *Social Value Judgment's* document was originally published in 2005 and was designed to provide guidance on new technologies. To accommodate the expansion of NICE's remit to include social care providers and local governance, a new list of principles was published most recently in 2020. The *Social Value Judgement* document from 2005 was formally replaced by the 2020 principles, but NICE has noted that the new list of principles builds on it and does not "signal a departure from our current methods and processes." See National Institute for Health and Care Excellence, "NICE Publishes Updated Principles," January 30, 2020, https://www.nice.org.uk/news/article/nice-publishes-updated-principles. The new principles can be found here: National Institute for Health and Care Excellence, "Our Principles," 2020, https://www.nice.org.uk/about/who-we-are/our-principles. For example, Principle 7 states that NICE's assessment is based on an assessment of population benefit and value for money, while Principle 9 states that one of the key aims is to reduce health inequalities. See NICE, "Our Principles." In 2012, NICE was renamed from the National Institute for Health and Clinical Excellence to the National Institute for Health and Care Excellence to reflect its new responsibilities for providing guidance to local governments and social-care providers.

[37] NICE, *Social Value Judgements*, 4.

[38] NICE, *Social Value Judgements*, 8.

- beneficence, and
- distributive justice.[39]

After listing these principles, the document points out that none of them are especially hard and fast. Autonomy, according to the NICE document, cannot "be applied universally or regardless of other social values," and neither the principles of non-maleficence nor beneficence will avoid the need to "balance the benefits and harms."[40] The widely accepted moral principles of respect for autonomy, non-maleficence, and beneficence will thus be applied in some cases but not in others, yet it is not clear how the required balancing will be done or who will do it.

The document's discussion of the fourth principle, distributive justice in healthcare, is more extensive. Noting that the "mismatch between demands and resources in healthcare leads to the problem of 'distributive justice,' or how to allocate limited healthcare resources fairly within society," the document observes that there exist, "broadly, two approaches that can be taken to resolve such problems in publicly funded healthcare systems": utilitarianism and egalitarianism.[41] A utilitarian allocation would allocate "resources to maximise the health of the community as a whole."[42] Yet the efficiency of utilitarianism can be "at the expense of fairness."[43] It "can allow the interests of minorities to be overridden by the majority; and it may not help in eradicating health inequities."[44] The egalitarian option, by contrast, distributes "resources to allow each individual to have a fair share. . . . It allows an adequate, but not necessarily maximum, level of healthcare, but raises questions as to what is 'fair.' "[45] In other words, there are two common approaches to addressing the issue of distributive justice, and each of them can lead to potentially undesirable or unpopular outcomes.

How does the document resolve this political and philosophical conundrum? The basic problem is that there "is no consensus as to which approach provides the more ethical basis for allocating resources."[46] Since there is no general agreement on how exactly to achieve distributive justice, the "Social Value Judgements" document focuses on transparency, with the goal being "procedural justice."[47] The point of seeking procedural justice is not to resolve

---

[39] See NICE, *Social Value Judgements*, 8, for definitions of these terms.
[40] NICE, *Social Value Judgements*, 8.
[41] NICE, *Social Value Judgements*, 9.
[42] NICE, *Social Value Judgements*, 9.
[43] NICE, *Social Value Judgements*, 9.
[44] NICE, *Social Value Judgements*, 9.
[45] NICE, *Social Value Judgements*, 9.
[46] NICE, *Social Value Judgements*, 9.
[47] In the *Social Value Judgements* documents, NICE cites, and relies heavily on, the work of two Harvard professors, Norman Daniels and James Sabin. See Norman Daniels and James E. Sabin, *Setting Limits*

conflicts between differing conceptions of distributive justice, but rather to ensure that "the processes by which healthcare decisions are reached are transparent, and that the reasons for the decisions are explicit."[48] It is not clear, however, that procedural justice always provides a sufficient basis for decision-making. One local U.K. health authority, wrestling with the fact that "the question of how scarce health-care resources should be allocated cannot be solved by specifying the criteria for a just process," concluded that the process needed to include "guidance from philosophical analysis in coming to decisions."[49] However those decisions are made, with or without the help of philosophy, procedural justice is designed to provide for "decision-makers to be 'accountable for their reasonableness.'"[50] This accountability requires NICE's decisions to exhibit "four characteristics": publicity, relevance ("The grounds for decisions must be ones that fair-minded people would agree are relevant"), challenge and revision of "decisions that are unreasonable," and some form of regulation to make sure the other three characteristics are present.[51]

Many of the difficult questions identified in the "Social Value Judgements" document have been troubling both philosophers and policymakers for a very long time. It would certainly be difficult to imagine the modern principle of non-maleficence arising without reference to Hippocrates's mandate to "do no harm," for example. Our societies have been discussing and debating these basic principles for millennia. Each new era raises additional issues concerning how to apply even the most basic and simple concepts, such as "do *no* harm," an injunction that is a simply impossible goal for physicians to follow to the letter in light of the side effects and rigors associated with even the highest-technology medicines. Scientists and physicians can figure out whether a new drug actually extends lives, and mathematicians can calculate the costs; but none of those analyses lead directly to a considered judgment about who should have those benefits and at what price.

Applying these issues to the question of how governments should allocate limited resources for healthcare is so difficult that the two Harvard scholars (a

Fairly: Learning to Share Resources for Health, 2nd ed. (New York: Oxford University Press, 2008), the first edition of which is referenced in NICE's report on social value judgments. See NICE, *Social Value Judgements*, 30n15.

[48] NICE, *Social Value Judgements*, 9–10. See also Daniels and Sabin, *Setting Limits*, 2.
[49] Tony Hope, John Reynolds, and Siân Griffiths, "Rationing Decisions: Integrating Cost-Effectiveness with Other Values," in *Medicine and Social Justice: Essays on the Distribution of Health Care*, ed. Rosamond Rhodes, Margaret P. Battin, and Anita Silvers (New York: Oxford University Press, 2002), 151 (describing the Oxfordshire Health Authority).
[50] NICE, *Social Value Judgements*, 10.
[51] NICE, *Social Value Judgements*, 10.

philosopher and a physician) who developed the concept of procedural justice in healthcare observe that "no democratic society we are aware of has achieved consensus on . . . distributive principles for healthcare."[52] In the end, the real decisions-makers will be the country's elected leaders, when fair-minded and not-so-fair-minded people declare the policymakers' recommendations to be unreasonable, unjust, or simply not to their liking.[53]

For the vast majority of U.K. citizens, the NHS's decision not to approve payment for a new drug is, practically speaking, the equivalent of the drug never having existed in the first place. Only a fairly small percentage of people will have sufficient personal wealth or private insurance to cover the high costs of new medicines if the NHS does not elect to pay for them. In fact, knowing that the drug exists but is inaccessible to patients covered by the NHS—while it is available to wealthier neighbors or to patients in more prosperous countries—may seem less fair and reasonable to the public than if the drug never existed at all. It will be cold comfort for patients covered by the NHS to know that the U.K.'s economy was stimulated by research funding that contributed to the development of a new drug if that stimulus was not financially potent enough to allow the nation to be able to afford to pay for the drug itself.

How these questions are answered by elected officials clearly has profound medical, moral, and political consequences. For our discussion of the future of the biotechnology industry, the same answers can have substantial and fairly immediate implications for the industry's economic health as well. NICE's recommendations currently influence about 30% of the global pharmaceutical market. If enough NHS-like organizations, or the politicians who decide how much funding to allocate to them, refuse to pay for these expensive new medicines, then the pharmaceutical companies will cease trying to create them, and they will stop acquiring them from biotech companies, which will become less attractive for venture capitalists and public investors. Shutting down the flow of private and public investment capital into these firms, in turn, will reduce or eliminate the number of new biotech companies. And as we saw in Chapter 7, fewer product development shots-on-goal by biotech companies will lead to considerably fewer new medicines.

Unless dramatic breakthroughs in life sciences research occur in ways that reduce the cost of new biomedical treatments by an order of magnitude or

---

[52] Daniels and Sabin, *Setting Limits*, 2. For a good summary of many—often conflicting—views on how principles of justice relate to healthcare, see the thirty-four essays from physicians, philosophers, lawyers, and public health professionals in Rhodes, Battin, and Silvers, *Medicine and Social Justice.*

[53] See Bernard H. Baumrin, who predicts that "competition to expand health-care services to meet health-care expectations will be the main social battlefield of the twenty-first century": Bernard H. Baumrin, "Why There Is No Right to Health Care," in Rhodes, Battin, and Silvers, *Medicine and Social Justice*, 83.

more—and despite many attempts by pharmaceutical and biotech companies to minimize costs, that shift does not appear to be on the scientific horizon—our economies will need to grow in other areas just to keep healthcare expenditures at their current level of GDP spending. (To see how these costs fit into the overall economic picture, it may be useful to note that, in 2019, the amount of GDP spent on medicines was 1.97% in Japan, 1.95% in the United States, and 1.23% in the United Kingdom.[54]) Alternatively, we will need to ration healthcare (or, in many countries, expand the rationing of healthcare) by setting a limit on overall healthcare expenditures, in which case there is likely to be a sharp reduction in investment capital available to an industry that is heavily reliant on the availability of that funding.

In sum, how policymakers and private payers signal to investors and companies that they will answer these questions in the future will have a dramatic effect on the willingness of private capital to invest in developing new drugs. The near-term economic growth stimulated by life sciences research and the biotech industry depends on credible evidence of a longer-term commitment—or at least a reason for private capital to believe in that potential commitment—to pay for the often expensive fruits of those labors. Failure by policymakers and payers in enough of the world's major pharmaceutical markets to show a reasonable degree of willingness to make those longer-term financial commitments will rapidly diminish both the economic growth that has been projected to be derived from discoveries in the life sciences and the number of medicines created to benefit the public that originally supported the basic research.

All Government All the Time?

Finally, since so much of this chapter has focused on the importance of private investment in healthcare research, it may be valuable to reflect on an entirely different scenario. The governments around the world now investing well over $100 billion per year in life-sciences research[55] could decide to "cut out the middleman" and essentially nationalize (or even multi-nationalize) pharmaceutical research. Instead of primarily supporting basic research and then looking to private companies and investment capital to provide the funding and expertise to develop, market, and sell the resulting drugs, governments

---

[54] See Organisation for Economic Co-operation and Development, "Pharmaceutical Spending (Indicator)," accessed September 29, 2020, https://data.oecd.org/healthres/pharmaceutical-spending.htm.
[55] See Justin Chakma et al., "Asia's Ascent—Global Trends in Biomedical R&D Expenditures," *New England Journal of Medicine* 370, no. 1 (January 2014): 4.

could become fully integrated drug developers. In fact, numerous programs sponsored by the National Institutes of Health and similar bodies elsewhere have increasingly been willing to support bench-to-bedside translational research. Eliminating pharmaceutical profits would have the potential to make medical costs lower, especially for high-margin, patent-protected products.

We believe that it is unlikely that it would be politically possible throughout the world to make it unlawful for for-profit companies to develop and commercialize pharmaceutical products, although enough governmental limitations on approvals or payments for new drugs could have virtually the same effect. And, of course, nothing currently prevents governments from developing drugs, either alone or in collaboration with academia or industry. In a number of cases, governmental research that has been licensed to for-profit companies has ultimately led to a successful product, and in the future, governments could decide to take a more active role in those further development efforts.

Whether governmental research in the absence of a significant effort by commercial entities would be a better, more efficient way to develop new drugs would be an intriguing topic to consider, but is well beyond the scope of this work other than to note that, based on the analysis in Chapter 7, any governmental efforts that have organizational structures and decision-making processes similar to those currently found in large pharmaceutical companies would be expected to be less innovative and less cost effective than those in the biotech industry. Additionally, at present, the pharmaceutical and biotechnology industries provide nearly 60% of the total amount of biomedical research funding in the United States.[56] This amount, over $70 billion per annum, would likely disappear in this scenario, along with hundreds of thousands of jobs. The government could certainly replace these positions in national laboratories, but doing so would involve an annual expense of many tens of billions of dollars, even if the new government scientists were to earn less than they did when they worked in private industry, since salaries are only a fraction of the cost of drug development.

Conclusion

In summary, then, the economic growth driven by the biotech industry over the last few decades has been fueled primarily by a potent combination of biomedical breakthroughs and a heavy infusion of private investment capital.

---

[56] See Chakma et al., "Asia's Ascent," 4.

For this capital-intensive industry to continue to generate new jobs and new drugs, society needs to pay high prices for those new medicines. Accordingly, especially in countries where the government funds both basic research in the life sciences *and* the purchase of new drugs, as we have seen in the United Kingdom (and increasingly in other countries), the biomedical discoveries need to be paid for twice, once at the beginning of the research enterprise and then again at its successful conclusion in the creation of a new treatment for disease. In between, the private sector will provide the funds to create the companies (and the new jobs) that will transform the basic biological research into therapeutic products.

With no molecular cost-saving solution to the cost of healthcare currently on the horizon, the industry's best bet may be to look for ways to make product development more efficient. Chapter 7 showed that it is not possible, based on current levels of knowledge in a highly ambiguous biomedical environment, to improve drug discovery by trying to identify and select future successes in advance. Rather than making large bets on relatively few opportunities that eventually do not pan out, the most efficient R&D strategy is placing smaller bets on many more possibilities. The highly decentralized nature of the biotech industry has allowed it to do precisely that, which is why it has outperformed the pharmaceutical industry in creating more breakthrough products at lower cost. Hence, a potential way to reduce the cost of new drugs is to shift more of the focus of biomedical R&D from large pharmaceutical companies to the biotech industry.

For the foreseeable future, the process of creating commercial blockbusters out of biomedical breakthroughs will continue to be primarily in the hands of the thousands of entrepreneurial entities that make up the biotech industry. To have the best chance of fulfilling that scientific and economic promise, the industry will need the continued support of investors who will be watching the decisions of public and private healthcare payers with great care.

A COVID-19 Manhattan Project?

By Donald Drakeman, Christoph Loch, and Nektarios Oraiopoulos[1]

What will it take to *quickly* find an effective COVID-19 treatment? Amid calls for a coronavirus czar or a new medical "Manhattan Project," we need to ask just what kind of process will be the most likely to succeed.

We have studied the origins of the highest priority new medicines of the last two decades—the ones that improve our treatment of patients with serious and life-threatening diseases—in other words, exactly what we need right now. The clear message is that a command-and-control process where the world's leading experts come together to focus resources on what seem to be the most promising ideas—an approach that would be the natural tendency of almost any czar—is *less* likely to come up with something new and important. Instead, what we need is a multiple-shots-on-goal approach that would allow us to test as many and as diverse ideas as possible, some of which will seem crazy to some experts, at least until they start working (or can unambiguously be ruled out).

For years, the diverse and uncoordinated efforts of small entrepreneurial biotech companies have outperformed the rigid structures often found in large, well-established companies in creating high-priority drugs, despite fewer resources and less experience. They have done so by trying many, many things; failing early and often; and eventually discovering genuinely novel and impressively effective medicines. Those successes are often completely unexpected. When the Nobel committee awarded the 2018 prize to James P. Allison of the University of California–Berkeley, it noted that his breakthrough in using the immune system to help combat cancer came "despite little interest from the pharmaceutical industry," and it was ultimately developed by a small biotech company.

The Manhattan Project, begun in 1939 to develop a nuclear bomb, is often perceived as the first modern example of a large-scale project that applied czar-like control-oriented principles (top-down planning, focus on hitting milestones, etc.), but it was far from that. In fact, knowing when to experiment widely and when to focus was the reason for its success. Initially expecting to manage a large-scale engineering project involved in assembling and deploying all the necessary resources to build the atomic bomb, General Leslie Groves soon learned that no one actually knew what was needed. There was no working design for the bomb, and there were widely disparate theories for how it should be designed.

Groves's team consisted of arguably the smartest group of scientists in the world. His vital insight was to take their various and conflicting ideas, and instead of trying to pick a winner, he pushed the candidates forward in parallel. That is, rather than ask for the consensus choice, he provided them with the resources to advance them all simultaneously, with the opportunity for them to learn from one another. He then focused on responding and adapting to unforeseen events as they occurred, including adding new candidate solutions mid-course, and the new

[1] Appendix 1 first appeared as an Insight article in the *California Management Review*: Donald Drakeman, Christoph Loch, and Nektarios Oraiopoulos, "A COVID-19 Manhattan Project?," *California Management Review*, May 22, 2020, https://cmr.berkeley.edu/2020/05/covid-manhattan-project/. The authors are grateful to Christoph Loch for graciously allowing us to include this essay in the book.

candidate (on plutonium enrichment) then rejuvenated a seemingly losing candidate (in bomb design). Doing so not only enhanced the overall chances of success but also sped up development in the midst of an extremely important race.

The (nowadays) conventional project management wisdom, of focusing on the "best" candidate (because that sounds efficient), makes sense only when you already know what success looks like. With a clear picture of all the required steps toward a successful outcome, carefully staged, sequential planning brings projects to completion on time and under budget.

But that is not where we are today with COVID-19 treatments. We do not yet know what characteristics a safe and effective treatment will have. In the current crisis, medical science has the broader remit of defeating the enemy by any and all means possible. We're more like Eisenhower than Groves. We don't yet know whether we want a bomb or a battleship or a missile. In fact, in the end, we probably will need a combination of treatments. We just need to win the war.

Medical science has responded to this new viral challenge with a burst of creative ideas. Around the world, university, government, and corporate scientists are repurposing old drugs, designing new ones, and trying to imagine what even more novel approaches might be effective in combating this virus. We need to cheer them *all* on.

What should a coronavirus czar do? First, protect the independence of all the various projects and resist the urge to centralize. Then do everything possible to make sure that each project at every university, government lab, and company gets what it asks for: access to samples, the cooperation of clinics for testing, collaborations with other labs with complementary technologies, and the removal of red tape. In short, make it easier and faster for every research effort, large and small.

Second, look for ways to facilitate learning across projects by encouraging as much open sharing of data, laboratory reagents, and results (both good and bad) as possible. Each time a question is answered, and our understanding of an area is improved, all teams need to know. For some teams this new knowledge might have little impact, but for others it might become a catalyst for their next set of questions. Collaboration should also be distinct from coordination. What we need is information sharing rather than goal alignment among the different teams. Digesting such information takes time, which is of essence amid the pandemic. But given that "winning" might entail different treatments or even combinations of different treatments for different patients, such learning might be essential.

Designing new drugs for life-threatening illnesses has become far more scientific over the past few decades of the biomedical revolution, but it has not yet become predictable. In an environment in which less than 10% of new drugs succeed—and we still cannot tell which ones in advance—any coronavirus czar or Manhattan Project–style leader needs to resist the temptation to place all our bets on whatever some group of experts thinks is most likely to succeed. Until the clinical data are in and we can see what will actually defeat the virus in infected patients, the key goal is to get the needed resources to the maximum number of projects and be ready to take advantage of the unknown and unforeseen as they develop. Then be prepared for a sudden shift to central command mode when something works and we need a hundred million doses as fast as humanly possible.

Bioscience researchers, and even many companies from the biotech and pharmaceutical industries, are collaborating to an unprecedented degree to defeat this virus. Coronavirus czars and political leaders around the world need to support this vital information sharing and encourage the spirit of open collaboration. We are all in this pandemic together, as COVID-19 knows no geographical borders.

Renovation as Innovation—Is Repurposing the Future of Drug Discovery Research?

By Arthur Neuberger, Nektarios Oraiopoulos, and
Donald L. Drakeman[1]

Repurposing is increasingly hailed as a solution to the problem of drug development's high costs and poor productivity. Instead of starting with a disease and beginning the laborious and expensive task of discovering a new drug, researchers can initiate the discovery process with a drug in hand and simply look for the right disease. The question is: Will repurposing fulfill its promise, and, if so, should we scale back our commitment to basic research and novel drug discovery and focus instead on second-chance drugs?

With repurposing success stories such as Viagra and thalidomide as inspiration, the National Institutes of Health offers grants to take advantage of failed drugs as "partially developed therapeutic candidates (assets)" that have "cleared several key steps along the development path," and can therefore "accelerat[e] the pace of . . . development."[2] The Food and Drug Administration (FDA) and the Medical Research Council have followed suit with repurposing initiatives, governments and foundations have offered grant funding, and companies have provided access to large libraries of compounds[3]—all with the goal of turning pharmaceutical failures into medical successes by testing unsuccessful compounds in different diseases.

Despite all the excitement—a new scientific paper with repurposing in the title appears nearly every day—there has been impressively little data on success rates. So far, repurposing advocates have told compelling stories about the well-known examples, but all of the numbers

[1] Appendix 2 was originally published as Arthur Neuberger, Nektarios Oraiopoulos, and Donald L. Drakeman, "Renovation as Innovation: Is Repurposing the Future of Drug Discovery Research?," *Drug Discovery Today* 24, no. 1 (January 2019): 1–3. The authors are grateful to Arthur Neuberger for graciously allowing us to include this essay in the book.

[2] "NCATS Announces Funding Opportunities to Repurpose Existing Drugs through Public-Private Partnerships," National Center for Advancing Translational Sciences (website), last modified February 2017, https://ncats.nih.gov/news/releases/2017/drug-repurpose-partnerships. See also Clare Thibodeaux, "New Repurposing Research Funding Opportunity for Rare Diseases," News/Events, Cures within Reach (website), last modified January 11, 2016, https://web.archive.org/web/20160518151029/http://cureswith inreach.org/newsroom/blog-re-rx/460-new-repurposing-research-funding-opportunity-for-rare-disea ses; National Center for Advancing Translational Sciences, *Discovering New Therapeutic Uses for Existing Molecules* (Bethesda, MD: National Institutes of Health, 2015), accessed online, https://ncats.nih.gov/files/ NTU-factsheet.pdf; "Biomedical Catalyst: Developmental Pathway Funding Scheme (DPFS) Outline: Mar 2021," UK Research and Innovation (website), last modified March 5, 2021, https://mrc.ukri.org/funding/ browse/biomedical-catalyst-dpfs/biomedical-catalyst-developmental-pathway-funding-scheme-dpfs-outline-mar-2021/; and Elie Dolgin, "Nonprofit Disease Groups Earmark Grants for Drug Repositioning," *Nature Medicine* 17, no. 9 (September 2011): 1027.

[3] See Mark S. Boguski, Kenneth D. Mandl, and Vikas P. Sukhatme, "Repurposing with a Difference," *Science* 324, no. 5933 (June 2009): 1394–95; Stephen M. Strittmatter, "Overcoming Drug Development Bottlenecks with Repurposing: Old Drugs Learn New Tricks," *Nature Medicine* 20, no. 6 (June 2014): 590–91; and Aidan Power, Adam C. Berger, and Geoffrey S. Ginsburg, "Genomics-Enabled Drug Repositioning and Repurposing: Insights from an IOM Roundtable Activity," *JAMA* 311, no. 20 (May 2014): 2063–64.

Table A2.1 Repurposing success rates

Repurposing	Same therapeutic area	Different therapeutic area
First success, $n = 167$	Attempts: (52/167) = 31% Successes: (35/52) = 67%	Attempts: (30/167) = 18% Successes: (10/30) = 33%
First failure, $n = 667$	Attempts: (104/667) = 16% Successes: (9/104) = 9%	Attempts: (65/667) = 10% Successes: (6/65) = 9%

Note: Repurposing rates ("Attempts") and corresponding commercial launch rates ("Success") in the same and different therapeutic area as the first attempted clinical development path for two subgroups: (1) first attempted clinical development path yielding a market launch ("First Success"); (2) first attempted clinical development path terminated ("First Failure").

have been estimates and projections. These estimates have ranged from 30% to as high as a potential 75% success rate.[4]

If such high success rates are possible, policymakers would need to rethink their long-standing commitments to basic biomedical research, where any successful outcomes will be far more time consuming, much more expensive, and considerably less likely. With thousands of partially developed compounds with which to work, and the FDA approving new molecular entities at the rate of only about twenty or so each year, a 75% (or even 30%) success rate for repurposing the thousands of failed drugs would satisfy the regulators' and payers' ability to absorb new medicines for a century or more.

In light of newly available data, repurposing certainly has promise, but it would be premature to abandon fundamental research in its favor. We have analyzed the complete clinical development history of 834 drug candidates that commenced clinical trials between 1980 and 2012. The analysis clearly shows that repurposing can be expected to yield important new medicines, but those successes are very unlikely to occur at the high projected rates (see Table A2.1).

In considering redirecting old drugs, it is important to distinguish between what may be the lower hanging fruit—that is, finding new uses for already approved drugs—and the harder-to-reach variety: turning failed drugs into successful ones in other diseases. Both "repurposing" and "repositioning" have been used to describe either or both of these phenomena, although most of the policy arguments for repurposing revolve around either failed drugs or successful ones that are redeveloped for entirely new therapeutic areas.

When it comes to evaluating the prospects for repurposing, there are two key questions to ask: (1) What are the chances that we will find a good reason to try a drug in a different disease from the one for which it was developed? (2) Once we have formulated that hypothesis, what is the likelihood of success?

Overall, 21% of the products we reviewed were approved by the FDA and launched commercially in the first indication for which they were tested. With a successful drug in hand, the pharmaceutical industry has regularly sought to expand its use in the same therapeutic area, such as testing a breast cancer drug in ovarian cancer.[5] Such line extensions were attempted for only 31% of the initial successes, but they had a high success rate: 67% of the products had at least one additional success in the same therapeutic area (cancer, autoimmune disease, cardiovascular disease, and so on) as the first indication.

[4] See Nicola Nosengo, "Can You Teach Old Drugs New Tricks?," *Nature* 534, no. 7607 (June 2016): 315.

[5] See Arthur G. Lipman, review of *Drug Repurposing and Repositioning: Workshop Summary*, by Sarah H. Beachy et al., *Journal of Pain & Palliative Care Pharmacotherapy* 29, no. 1 (February 2015): 81.

More relevant to the repurposing question is how these successful drugs fared in different therapeutic areas. A considerably smaller percentage of these initial successes entered clinical testing in a different therapeutic area, and the success rate dropped dramatically. Only 18% of the products were tried in another area, and only 33% of those attempts succeeded.

The alchemy involved in turning failures into successes is even more complicated. Only 16% were tried in another indication, and most were in the same therapeutic area, with only 10% of the initial failures tried in a different area altogether. Whether the therapeutic area was the same or different, the success rate was the same: 9%.

Despite the difficulty of finding a good hypothesis in these classic repurposing cases and then proving it with clinical results, the rewards can be substantial. Some of these once-failed drugs have become very valuable commercial and medical products. Two of them, Remicade and Enbrel, both initially developed for treating sepsis, are approved for the treatment of rheumatoid arthritis and were ranked among the top five best-selling drugs in 2016.[6]

Interestingly, and adding to the challenges presented by repurposing, we could find no predictive pattern among the successfully repurposed products. The success rates for small and large molecules were similar, and the successes were distributed across a range of therapeutic areas. The only pattern that did emerge clearly was that most of the attempts and most of the successes were in the same therapeutic area, which is outside of repurposing's main theme of finding entirely new uses for existing molecules.

How promising are these success rates? Despite ever more sophisticated methods for conducting pharmaceutical research, clinical success rates have remained disappointingly low. The most recent major study[7] shows about a 10% success rate for all of the therapeutic indications for which potential new medicines have entered clinical trials. Additionally, various publications have cited a 5% success rate from the start of formal pre-IND toxicology testing.[8]

These low rates of success have been accompanied by inexorably increasing drug development costs, a phenomenon sometimes called "Eroom's law," which says that the "efficiency of research and development of new drugs . . . halves every nine years or so . . . Moore's law for microprocessors in reverse."[9] In this environment, it is not surprising that the promise of repurposing partially developed drugs has generated considerable enthusiasm from patients, payers, regulators, policymakers, and biopharmaceutical companies.

Based on our new data, does repurposing deserve all the attention it has been getting? To be sure, initial successes are a reasonably good indicator of successful follow-ons, and initial failures do not necessarily signal the ultimate failure of the product. Accordingly, for those products for which large amounts of time and resources have been invested in research, development, large-scale manufacturing, and clinical testing, there can be good reasons to consider the possibility of repurposing, even after an initial clinical failure.

At the same time, turning base metal into gold has been difficult for a long time, and repurposing is unlikely to fill our future medicine cabinets single-handedly, especially if we focus on only products that we hope will be effective in another disease setting. Of the 834 new molecules entering clinical trials, less than 2% were ultimately launched in a therapeutic area

[6] See Alex Philippidis, "The Top 15 Best-Selling Drugs of 2016," *GEN: Genetic Engineering & Biotechnology News*, March 6, 2017, https://www.genengnews.com/the-lists/the-top-15-best-selling-drugs-of-2016/77900868.

[7] Michael Hay et al., "Clinical Development Success Rates for Investigational Drugs," *Nature Biotechnology* 32, no. 1 (January 2014): 47.

[8] See, for example, Alexander Schuhmacher, Oliver Gassmann, and Markus Hinder, "Changing R&D Models in Research-based Pharmaceutical Companies," *Journal of Translational Medicine* 14, no. 105 (April 2016): 2.

[9] Nosengo, "New Tricks," 316.

other than the one in which they were originally tested. The vast majority of the new medicines were the products of de novo discovery that succeeded in the originally expected therapeutic area. Accordingly, as promising as repurposing may be, basic research remains essential. Not only will it help us understand where failed drugs might have a second chance but, more importantly, it will continue to create a fertile environment for the often unexpected discoveries that can launch entirely new classes of medicines.

Bibliography

Abbott, Alison. "Genomics: Sorry, Dogs—Man's Got a New Best Friend." *Nature* 420, no. 6917 (December 2002): 729.

Adams, Ben. "Biotechs Getting Bigger in Late-Stage R&D, Leaving Big Pharmas Behind: Report." *Fierce Biotech*, April 23, 2019. https://www.fiercebiotech.com/biotech/biotechs-getting-bigger-late-stage-r-d-leaving-big-pharmas-behind-report.

Adams, Ben. "CRO Market to Recover, Using 'Hybrid trials,' with Revenue Hitting $64B by 2024: Report." *FiercePharma,* September 8, 2020. https://www.fiercebiotech.com/cro/cro-market-to-recover-using-hybrid-trials-revenue-hitting-64b-by-2024-report.

Adegbesan, J. Adetunji, and Matthew J. Higgins. "The Intra-alliance Division of Value Created through Collaboration." *Strategic Management Journal* 32, no. 2 (February 2011): 187–211.

Aghion, Philippe, and Jean Tirole. "The Management of Innovation." *Quarterly Journal of Economics* 109, no. 4 (November 1994): 1185–209.

Agrawal, Vishal, and Nektarios Oraiopoulos. "The Role of Decision Rights in Codevelopment Initiatives." *Manufacturing & Service Operations Management* 22, no. 4 (July–August 2020): 832–49.

"Amgen and J&J Fight It Out over Procrit." *The Pharma Letter*, January 20, 2000. https://www.thepharmaletter.com/article/amgen-and-j-j-fight-it-out-over-procrit.

Anand, Jaideep, Raffaele Oriani, and Roberto S. Vassolo. "Alliance Activity as a Dynamic Capability in the Face of a Discontinuous Technological Change." *Organization Science* 21, no. 6 (November/December 2010): 1213–32.

Antelo, Manel. "Licensing a Non-Drastic Innovation under Double Information Asymmetry." *Research Policy* 32, no. 3 (March 2003): 367–90.

Argenx. "Argenx Raises $750 Million in Gross Proceeds in a Global Offering." News. Last modified May 28, 2020. https://www.argenx.com/news/argenx-raises-750-million-gross-proceeds-global-offering.

Arora, Ashish, and Alfonso Gambardella. "Complementarity and External Linkages: The Strategies of the Large Firms in Biotechnology." *Journal of Industrial Economics* 38, no. 4 (June 1990): 361–79.

Arora, Ashish, Alfonso Gambardella, Laura Magazzini, and Fabio Pammolli. "A Breath of Fresh Air? Firm Type, Scale, Scope, and Selection Effects in Drug Development." *Management Science* 55, no. 10 (October 2009): 1638–53.

Association of University Technology Managers. *AUTM Licensing Survey: FY 1996.* Edited by Daniel E. Massing. Norwalk, CT: AUTM, 1997.

Association of University Technology Managers. *AUTM Licensing Survey: FY 1997.* Edited by Daniel E. Massing. Norwalk, CT: AUTM, 1998.

Association of University Technology Managers. *AUTM Licensing Survey: FY 2000,* edited by Lori Pressman. Northbrook, IL: AUTM, 2002.

Association of University Technology Managers. *2012 AUTM U.S. Licensing Activity Appendix.* Deerfield, IL: AUTM, 2013.

Association of University Technology Managers. *The AUTM Briefing Book: 2015.* Deerfield, IL: AUTM, 2015. https://www.cshl.edu/wp-content/uploads/2017/12/AUTM-Briefing-Book-2015.pdf.

Audretsch, David B., and Maryann P. Feldman. "Small-Firm Strategic Research Partnerships: The Case of Biotechnology." *Technology Analysis & Strategic Management* 15, no. 2 (June 2003): 273–88.

Baker, Laurence, and Bruce Deal. *Economic Impact Analysis: Proposition 71 California Stem Cell Research and Cures Initiative.* Boston: Analysis Group, 2004. https://www.analysisgroup. com/uploadedfiles/content/news_and_events/news/proposition_71_report.pdf.

Battelle and Biotechnology Industry Organization. *Battelle/BIO State Bioscience Jobs, Investments and Innovation 2014.* Washington, DC: Biotechnology Industry Organization, 2014. https://www.bio.org/sites/default/files/legacy/bioorg/docs/files/Battelle-BIO-2014-Industry.pdf.

Battelle Technology Partnership Practice. *Biopharmaceutical Industry-Sponsored Clinical Trials: Impact on State Economies.* March 2015. http://phrma-docs.phrma.org/sites/default/ files/pdf/biopharmaceutical-industry-sponsored-clinical-trials-impact-on-state-econom ies.pdf.

Baumrin, Bernard H. "Why There Is No Right to Health Care." In *Medicine and Social Justice: Essays on the Distribution of Health Care,* edited by Rosamond Rhodes, Margaret Battin, and Anita Silvers 78–83. New York: Oxford University Press, 2002.

Bayh–Dole 40. Accessed September 1, 2020. https://bayhdole40.org.

Bazell, Robert. *Her-2: The Making of Herceptin, a Revolutionary Treatment for Breast Cancer.* New York: Random House, 1998.

Beg, Sarwar, Mahfoozur Rahman, Syed Sarim Imam, Nabil K. Alruwaili, Majed Al Robaian, and Sunil Kumar Panda, eds. *Pharmaceutical Drug Product Development and Process Optimization: Effective Use of Quality by Design.* Palm Bay, FL: Apple Academic Press, 2020.

Behnke, Nils, and Norbert Hueltenschmidt. *Changing Pharma's Innovation DNA.* Boston: Bain, 2010. https://media.bain.com/Images/BAIN_BRIEF_Changing_pharmas_innovation_ DNA.pdf.

Bekelman, Justin E., Yan Li, and Cary P. Gross. "Scope and Impact of Financial Conflicts of Interest in Biomedical Research: A Systematic Review." *Journal of the American Medical Association* 289, no. 4 (January 2003): 454–65.

Binder, Gordon, and Philip Bashe. *Science Lessons: What the Business of Biotech Taught Me about Management.* Cambridge, MA: Harvard Business School Press, 2008.

Biomart. "Market and R&D Analysis of Recombinant Protein Drugs." *Creative Biomart Blog,* June 9, 2017. https://www.creativebiomart.net/blog/market-and-rd-analysis-of-recombin ant-protein-drugs/.

Biotech and Money. "Interview: Killer Experiments and Keys to a Successful Deal." *Drugs & Dealers Magazine,* January 2015. http://cdn2.hubspot.net/hub/378634/file-2301474614-pdf/DDJan2015PDF.pdf.

Biyalogorsky, Eyal, William Boulding, and Richard Staelin. "Stuck in the Past: Why Managers Persist with New Product Failures." *Journal of Marketing* 70, no. 2 (April 2006): 108–21.

Blankenship, Kyle. "The Top 20 Drugs by Global Sales in 2019." *FiercePharma,* July 27, 2020. https://www.fiercepharma.com/special-report/top-20-drugs-by-global-sales-2019.

Blumenthal, David. "Academic–Industrial Relationships in the Life Sciences." *New England Journal of Medicine* 349, no. 25 (December 2003): 2452–59.

Blumenthal, David, Nancyanne Causino, Eric Campbell, and Karen Seashore Louis. "Relationships between Academic Institutions and Industry in the Life Sciences—an Industry Survey." *New England Journal of Medicine* 334, no. 6 (February 8, 1996): 368–73.

Boguski, Mark S., Kenneth D. Mandl, and Vikas P. Sukhatme. "Repurposing with a Difference." *Science* 324, no. 5933 (June 2009): 1394–95.

Bolten, Barbara. "Fastest Drug Developers and Their Practices." *CenterWatch,* August 1, 2017. https://www.centerwatch.com/articles/13284.

Booth, Bruce. "Correlation's Fresh Look at Venture Capital Returns." *Forbes*, November 18, 2013. https://www.forbes.com/sites/brucebooth/2013/11/18/correlations-fresh-look-at-venture-capital-returns/#7af4a0e040e5.

Booth, Bruce. "New Blood Needed: Pharma R&D Leadership Tenure." *Forbes*, September 5, 2017. https://www.forbes.com/sites/brucebooth/2017/09/05/new-blood-needed-pharma-rd-leadership-tenure/#217ccc7e409a.

Booth, Bruce. "Our Experience with First-Time Biotech CEOs: Five Behaviors That Matter." *Life Sci VC* (blog), May 25, 2016. https://lifescivc.com/2016/05/experience-first-time-biotech-ceos-five-behaviors-matter/.

Booth, Bruce L., and Bijan Salehizadeh. "In Defense of Life Sciences Venture Investing." *Nature Biotechnology* 29, no. 7 (July 2011): 579–83.

Boudreau, Kevin J., Eva C. Guinan, Karim R. Lakhani, and Christoph Riedl. "Looking across and Looking beyond the Knowledge Frontier: Intellectual Distance, Novelty, and Resource Allocation in Science." *Management Science* 62, no. 10 (October 2016): 2765–83.

Brown, Shona L., and Kathleen M. Eisenhardt. "Product Development: Past Research, Present Findings, and Future Directions." *Academy of Management Review* 20, no. 2 (April 1995): 343–78.

Brynner, Rock, and Trent Stephens. *Dark Remedy: The Impact of Thalidomide and Its Revival as a Vital Medicine*. New York: Basic Books, 2001.

Bryson, Bill. *A Short History of Nearly Everything*. New York: Broadway Books, 2004.

Bussgang, Jeffrey. "Are You Suited for a Start-Up?" *Harvard Business Review* 95, no. 6 (November/December 2017): 150–53.

Cambridge Judge Business School. "Dr. Lisa Drakeman, Former CEO of Genmab: Entrepreneurs Must Accept Rejection but Keep Going." News & Insight. Last modified June 18, 2013. https://www.jbs.cam.ac.uk/insight/2013/dr-lisa-drakeman-former-ceo-of-genmab-entrepreneurs-must-accept-rejection-but-keep-going/.

Carreyrou, John. *Bad Blood: Secrets and Lies in a Silicon Valley Startup*. New York: Knopf, 2018.

Carroll, John. "Big Pharma Is Using Its Venture Cash to Outsource Early R&D to Biotech." *Fierce Biotech*, July 31, 2014. https://www.fiercebiotech.com/venture-capital/big-pharma-using-its-venture-cash-to-outsource-early-r-d-to-biotech.

Carroll, John. "GSK Déjà Vu: Time for a Top-Down Switch as New CEO Tries to Conquer Old R&D Demons." *Endpoints News*, August 7, 2017. https://endpts.com/gsk-deja-vu-time-for-a-top-down-switch-as-new-ceo-tries-to-conquer-its-old-rd-demons/.

Cassar, Gavin, and Henry Friedman. "Does Self-Efficacy Affect Entrepreneurial Investment?" *Strategic Entrepreneurship Journal* 3, no. 3 (September 2009): 241–60.

"Celera Sees Slightly Stronger Revenue, Narrowed Loss in Q4," *GenomeWeb*, July 25, 2002, https://www.genomeweb.com/archive/celera-sees-slightly-stronger-revenue-narrowed-loss-q4.Celgene. "A Pipeline of Innovative Therapies to Improve Patient Care." Newsroom. Last modified January 6, 2014. https://www.celgene.com/pipeline-committed-to-patients/.

Chakma, Justin, Gordon H. Sun, Jeffrey D. Steinberg, Stephen M. Sammut, and Reshma Jagsi. "Asia's Ascent—Global Trends in Biomedical R&D Expenditures." *New England Journal of Medicine* 370, no. 1 (January 2014): 3–6.

Chan, Su Han, John W. Kensinger, Arthur J. Keown, and John D. Martin. "Do Strategic Alliances Create Value?" *Journal of Financial Economics* 46, no. 2 (November 1997): 199–221.

Chinwalla, Asif T., et al. "Initial Sequencing and Comparative Analysis of the Mouse Genome." *Nature* 420, no. 6915 (December 2002): 520–62.

Christensen, Clayton M. *The Innovator's Dilemma: When New Technologies Cause Great Firms to Fail*. Boston: Harvard Business School Press, 2016.

Coff, Russell W. "How Buyers Cope with Uncertainty When Acquiring Firms in Knowledge-Intensive Industries: Caveat Emptor." *Organization Science* 10, no. 2 (March/April 1999): 144–61.

Cohen, Hal. "A Biotech by Any Other Name." *The Scientist* 17, no. 10 (May 2003). https://www.the-scientist.com/profession/a-biotech-by-any-other-name-51640.

Colaizzi, John L. *New Jersey: The Nation's Medicine Chest*. Nutley, NJ: Hoffmann-La Roche, 1985.

Collins, Christopher J., Paul J. Hanges, and Edwin A. Locke. "The Relationship of Achievement Motivation to Entrepreneurial Behavior: A Meta-analysis." *Human Performance* 17, no. 1 (2004): 95–117.

Criscuolo, Paola, Linus Dahlander, Thorsten Grohsjean, and Ammon Salter. "Evaluating Novelty: The Role of Panels in the Selection of R&D Projects." *Academy of Management Journal* 60, no. 2 (April 2017): 433–60.

Crommelin, Daan J. A., Robert D. Sindelar, and Bernd Meibohm, eds. *Pharmaceutical Biotechnology: Fundamentals and Applications*. 5th ed. New York: Springer, 2019.

Crook, Stanley T. "Progress in Antisense Technology." *Annual Review of Medicine* 55 (2004): 61–95.

Csaszar, Felipe A. "Organizational Structure as a Determinant of Performance: Evidence from Mutual Funds." *Strategic Management Journal* 33, no. 6 (June 2012): 611–32.

Dale, David C. "The Discovery, Development and Clinical Applications of Granulocyte Colony-Stimulating Factor." *Transactions of the American Clinical and Climatological Association* 109 (1998): 27–38.

Daniels, Norman, and James E. Sabin. *Setting Limits Fairly: Learning to Share Resources for Health*. 2nd ed. New York: Oxford University Press, 2008.

Danzon, Patricia M., and Sean Nicholson, eds. *The Oxford Handbook of the Economics of the Biopharmaceutical Industry*. New York: Oxford University Press, 2012.

Danzon, Patricia M., Sean Nicholson, and Nuno Sousa Pereira. "Productivity in Pharmaceutical–Biotechnology R&D: The Role of Experience and Alliances." *Journal of Health Economics* 24, no. 2 (March 2005): 317–39.

Darby, Michael R., and Lynne G. Zucker. "Going Public When You Can in Biotechnology." NBER Working Paper Series 8954. Cambridge, MA: National Bureau of Economic Research, May 2002. https://www-nber-org.proxy.library.upenn.edu/papers/w8954.pdf.

Davies, Kevin, John Russel, and John Dodge. "John Craig Venter Unvarnished." *Bio-IT World*, November 12, 2002. https://web.archive.org/web/20021212085635/http://www.bio-itworld.com/archive/111202/horizons_venter.html.

Davies, Kevin, and Michael White. *Breakthrough: The Race to Find the Breast Cancer Gene*. New York: John Wiley & Sons, 1996.

Davis, Caroline. "UK Poor Cousins to US Tech Transfers." *Times Higher Education*, June 6, 2003. https://www.timeshighereducation.com/news/uk-poor-cousins-to-us-tech-transfers/177247.article.

de Rond, Mark. *Strategic Alliances as Social Facts: Business, Biotechnology & Intellectual History*. Cambridge: Cambridge University Press, 2003.

Dessain, Scott, and Scott Fishman. *Preserving the Promise: Improving the Culture of Biotech Investments*. London: Academic Press, 2017.

Dewatripont, M., and E. Maskin. "Credit and Efficiency in Centralized and Decentralized Economies." *Review of Economic Studies* 62, no. 4 (October 1995): 541–55.

DiMasi, J. A., L. Feldman, A. Seckler, and A. Wilson. "Trends in Risks Associated with New Drug Development: Success Rates for Investigational Drugs." *Clinical Pharmacology & Therapeutics* 87, no. 3 (March 2010): 272–77.

DiMasi, Joseph A., Henry G. Grabowski, and Ronald W. Hansen. "Innovation in the Pharmaceutical Industry: New Estimates of R&D Costs." *Journal of Health Economics* 47 (May 2016): 20–33.

DiMasi, Joseph A., Ronald W. Hansen, and Henry G. Grabowski. "The Price of Innovation: New Estimates of Drug Development Costs." *Journal of Health Economics* 22, no. 2 (March 2003): 151–85.

DiMasi, Joseph A., Ronald W. Hansen, Henry G. Grabowski, and Louis Lasagna. "Cost of Innovation in the Pharmaceutical Industry." *Journal of Health Economics* 10, no. 2 (July 1991): 107–42.

Ding, Waverly W. "The Impact of Founders' Professional-Education Background on the Adoption of Open Science by For-Profit Biotechnology Firms." *Management Science* 57, no. 2 (February 2011): 257–73.

Dolgin, Elie. "Nonprofit Disease Groups Earmark Grants for Drug Repositioning." *Nature Medicine* 17, no. 9 (September 2011): 1027.

Dorey, Emma. "GlaxoSmithKline Presents a Biotech Façade." *Nature Biotechnology* 19, no. 4 (April 2001): 294–95.

Dougherty, Deborah, and Cynthia Hardy. "Sustained Product Innovation in Large, Mature Organizations: Overcoming Innovation-to-Organization Problems." *Academy of Management Journal* 39, no. 5 (October 1996): 1120–53.

Drakeman, Donald L. Review of *The Billion Dollar Molecule: One Company's Quest for the Perfect Drug,* by Barry Werth. *U.S.* 1 (August 1994): 49.

Drakeman, Donald L. "Benchmarking Biotech and Pharmaceutical Product Development." *Nature Biotechnology* 32, no. 7 (July 2014): 621–25.

Drakeman, Donald L. *Why We Need the Humanities: Life Science, Law and the Common Good.* New York: Palgrave Macmillan, 2016.

Drakeman, Donald, Christoph Loch, and Nektarios Oraiopoulos. "A COVID-19 Manhattan Project?" *California Management Review*, May 22, 2020. https://cmr.berkeley.edu/2020/05/covid-manhattan-project/.

Drakeman, Donald, and Nektarios Oraiopoulos. "The Risk of De-Risking Innovation: Optimal R&D Strategies in Ambiguous Environments." *California Management Review* 62, no. 3 (May 2020): 42–63.

Drews, Juergen. "Genomic Sciences and the Medicine of Tomorrow." *Nature Biotechnology* 14, no. 11 (November 1996): 1516–18.

Drivas, Kyriakos, Zhen Lei, and Brian Wright. *A Preliminary View of University of California Data on Disclosures, Licensing and Patenting.* Washington, DC: National Academy of Sciences, 2009. https://sites.nationalacademies.org/cs/groups/pgasite/documents/webpage/pga_058895.pdf.

Druker, Brian J., Moshe Talpaz, Debra J. Resta, Bin Peng, Elisabeth Buchdunger, John M. Ford, Nicholas B. Lydon, et al. "Efficacy and Safety of a Specific Inhibitor of the BCR-ABL Tyrosine Kinase in Chronic Myeloid Leukemia." *New England Journal of Medicine* 344, no. 14 (April 2001): 1031–37.

Dushnitsky, Gary, and Michael J. Lenox. "When Do Incumbents Learn from Entrepreneurial Ventures? Corporate Venture Capital and Investing Firm Innovation Rates." *Research Policy* 34, no. 5 (June 2005): 615–39.

Ellsberg, Daniel. "Risk, Ambiguity, and the Savage Axioms." *Quarterly Journal of Economics* 75, no. 4 (November 1961): 643–69.

Ernst & Young. *Beyond Borders: Global Biotechnology Report 2009.* London: Ernst & Young, 2009.

Ernst & Young. *Beyond Borders: Matters of Evidence—Biotechnology Industry Report 2013.* London: Ernst & Young, 2013.

Ernst & Young. *Beyond Borders: Unlocking Value—Biotechnology Industry Report 2014.* London: Ernst & Young, 2014.

Ernst & Young. *Beyond Borders: Reaching New Heights—Biotechnology Industry Report 2015.* London: Ernst & Young, 2015.

Ernst & Young. *Beyond Borders: Returning to Earth—Biotechnology Report 2016.* London: Ernst & Young, 2016.

Ernst & Young. *Beyond Borders: Staying the Course—Biotechnology Industry Report 2017*. London: Ernst & Young, 2017.

Ernst & Young. *How Will Deals Done Now Deliver What the Health Ecosystem Needs Next? 2020 EY M&A Firepower Report*. London: Ernst & Young, 2020. https://assets.ey.com/content/dam/ey-sites/ey-com/en_gl/topics/life-sciences/life-sciences-pdfs/ey-firepower-report-2020-how-will-deals-done-now-deliver-what-the-health-ecosystem-needs-next-v2.pdf.

Evaluate. *EvaluatePharma World Preview 2019, Outlook to 2024*. June 2019. https://www.evaluate.com/thought-leadership/pharma/evaluatepharma-world-preview-2019-outlook-2024.

FDAnews. "NICE Turns Down Roche's Polivy for Lymphoma Combination." Last modified March 5, 2020. https://www.fdanews.com/articles/196131-nice-turns-down-roches-polivy-for-lymphoma-combination.

Fraser, Laura. "Genentech Goes Public." Genentech, April 28, 2016. https://www.gene.com/stories/genentech-goes-public.

Friedman, Lawrence M., Curt D. Furberg, David L. DeMets, David M. Reboussin, and Christopher B. Granger. *Fundamentals of Clinical Trials*. 5th ed. New York: Springer, 2015.

Gallini, Nancy T., and Brian D. Wright. "Technology Transfer under Asymmetric Information." *RAND Journal of Economics* 21, no. 1 (Spring 1990): 147–60.

Ganti, Akhilesh. "Median." *Investopedia*, August 22, 2019. https://www.investopedia.com/terms/m/median.asp.

Geiger, Roger L., and Cresco M. Sá. *Tapping the Riches of Science: Universities and the Promise of Economic Growth*. Cambridge, MA: Harvard University Press, 2009.

Ginn, Samantha L., Anais K. Amaya, Ian E. Alexander, Michael Edelstein, and Mohammad R. Abedi. "Gene Therapy Clinical Trials Worldwide to 2017: An Update." *Journal of Gene Medicine* 20, no. 5 (March 2018): 1–16.

Glasure, Elizabeth. "BioSpace Feature: A Look at Miracle Drug Humira's Journey to Proven Efficacy." *BioSpace*, December 5, 2018. https://www.biospace.com/article/biospace-feature-a-look-at-miracle-drug-humira-s-journey-to-proven-efficacy-/.

"Global Monoclonal Antibodies (mAbs) Market Report 2020 with Profiles of Johnson & Johnson, Merck, AbbVie, Amgen, GlaxoSmithKline—ResearchAndMarkets.com." *Business Wire*, December 11, 2019. https://www.businesswire.com/news/home/20191211005627/en/Global-Monoclonal-Antibodies-mAbs-Market-Report-2020.

Godfrey, Paul C., Gove N. Allen, and David Benson. "The Biotech Living and the Walking Dead." *Nature Biotechnology* 38, no. 2 (January 2020): 132–41.

Gompers, Paul A. "Optimal Investment, Monitoring, and the Staging of Venture Capital." *Journal of Finance* 50, no. 5 (December 1995): 1461–89.

Gompers, Paul A. "Grandstanding in the Venture Capital Industry." *Journal of Financial Economics* 42, no. 1 (September 1996): 133–56.

Gompers, Paul A., Will Gornall, Steven N. Kaplan, and Ilya A. Strebulaev. "How Do Venture Capitalists Make Decisions?" *Journal of Financial Economics* 135, no. 1 (January 2020): 169–90.

Gompers, Paul, Anna Kovner, Josh Lerner, and David Scharfstein. "Venture Capital Investment Cycles: The Impact of Public Markets." *Journal of Financial Economics* 87, no. 1 (January 2008): 1–23.

Gompers, Paul, and Josh Lerner. "What Drives Venture Capital Fundraising?" *Brookings Papers on Economic Activity. Microeconomics* 29, no. 1998 (July 1998): 149–204.

Gompers, Paul, and Josh Lerner. "The Determinants of Corporate Venture Capital Success: Organizational Structure, Incentives, and Complementarities." In *Concentrated Corporate Ownership*, edited by Randall K. Morck, 17–54. Chicago: University of Chicago Press, 2000.

Goodman, Ellen. "Medicine Needs More 'Chumps.'" *Boston Globe*, March 1, 2001, A15.

Gordon, Siamon. "Phagocytosis: The Legacy of Metchnikoff." *Cell* 166, no. 5 (August 2016): 1065–68.

Gorman, Michael, and William A. Sahlman. "What Do Venture Capitalists Do?" *Journal of Business Venturing* 4, no. 4 (July 1989): 231–48.

Graham, Bradley. "Patent Bill Seeks Shift to Bolster Innovation." *Washington Post*, April 8, 1979. https://www.washingtonpost.com/archive/business/1979/04/08/patent-bill-seeks-shift-to-bolster-innovation/db14f277-ec0e-4ca5-9aeb-ce2cad86e25b/.

Grossman, Sanford J., and Oliver D. Hart. "The Costs and Benefits of Ownership: A Theory of Vertical and Lateral Integration." *Journal of Political Economy* 94, no. 4 (August 1986): 691–719.

Gulati, Ranjay. "Alliances and Networks." *Strategic Management Journal* 19, no. 4 (April 1998): 293–317.

Guo, Chunqing, Masoud H. Manjili, John R. Subjeck, Devanand Sarkar, Paul B. Fisher, and Xiang-Yang Wang. "Therapeutic Cancer Vaccines: Past, Present and Future." *Advances in Cancer Research* 199 (2013): 421–75.

Gyngell, Christopher, Thomas Douglas, and Julian Savulescu. "The Ethics of Germline Gene Editing." *Journal of Applied Philosophy* 34, no. 4 (August 2017): 498–513.

Hall, Robert E., and Susan E. Woodward. "The Burden of the Nondiversifiable Risk of Entrepreneurship." *American Economic Review* 100, no. 3 (June 2010): 1163–94.

Hancock, Diana, David B. Humphrey, and James A. Wilcox. "Cost Reductions in Electronic Payments: The Roles of Consolidation, Economies of Scale, and Technical Change." *Journal of Banking & Finance* 23, nos. 2–4 (February 1999): 391–421.

Hart, David M., ed. *The Emergence of Entrepreneurship Policy: Governance, Start-Ups, and Growth in the US Knowledge Economy.* New York: Cambridge University Press, 2003.

Hart, Oliver, and John Moore. "Property Rights and the Nature of the Firm." *Journal of Political Economy* 98, no. 6 (December 1990): 1119–58.

Hartwell, Leland H., Michael L. Goldberg, Janice A. Fischer, Leroy Hood, and Charles F. Aquadro. *Genetics: From Genes to Genomes.* 5th ed. New York: McGraw Hill, 2014.

Hawkins, John. *Conscience and Courage: How Visionary CEO Henri Termeer Built a Biotech Giant and Pioneered the Rare Disease Industry.* Cold Springs Harbor, NY: Cold Springs Harbor Laboratory Press, 2019.

Hay, Michael, David W. Thomas, John L. Craighead, Celia Economides, and Jesse Rosenthal. "Clinical Development Success Rates for Investigational Drugs." *Nature Biotechnology* 32, no. 1 (January 2014): 40–51.

Health Economics Research Group (HERG) Brunel University, Office of Health Economics (OHE), and RAND Europe. *Medical Research: What's It Worth? Estimating the Economic Benefits from Medical Research in the UK.* London: UK Evaluation Forum, 2008. https://mrc.ukri.org/publications/browse/medical-research-whats-it-worth/.

Hellmann, Thomas, and Manju Puri. "Venture Capital and the Professionalization of Start-Up Firms: Empirical Evidence." *Journal of Finance* 57, no. 1 (February 2002): 169–97.

Herper, Matthew. "The Cost of Developing Drugs Is Insane. That Paper That Says Otherwise Is Insanely Bad." *Forbes*, October 16, 2017. https://www.forbes.com/sites/matthewherper/2017/10/16/the-cost-of-developing-drugs-is-insane-a-paper-that-argued-otherwise-was-insanely-bad/#ac88ceb2d459.

Herper, Matthew. "The World's Best-Selling Drugs." *Forbes*, March 16, 2004. https://www.forbes.com/2004/03/16/cx_mh_0316bestselling.html#7bd44e5e8e38.

Higgins, Matthew J. "The Allocation of Control Rights in Pharmaceutical Alliances." *Journal of Corporate Finance* 13, no. 1 (March 2007): 58–75.

Higgins, Matthew J., and Daniel Rodriguez. "The Outsourcing of R&D through Acquisitions in the Pharmaceutical Industry." *Journal of Financial Economics* 80, no. 2 (May 2006): 351–83.

Highsmith, Jim. *Agile Project Management: Creating Innovative Products*. 2nd ed. Upper Saddle River, NJ: Addison-Wesley, 2009.

Hine, Damian, and John Kapeleris. *Innovation and Entrepreneurship in Biotechnology, an International Perspective: Concepts, Theories and Cases*. Cheltenham, UK: Edward Elgar, 2006.

Hope, Tony, John Reynolds, and Siân Griffiths. "Rationing Decisions: Integrating Cost-Effectiveness with Other Values." In Rhodes, Battin, and Silvers, *Medicine and Social Justice: Essays on the Distribution of Health Care*, edited by Rosamond Rhodes, Margaret Battin, and Anita Silvers, 144–55. New York: Oxford University Press, 2002.

Hsu, David H. "Venture Capitalists and Cooperative Start-Up Commercialization Strategy." *Management Science* 52, no. 2 (February 2006): 204–19.

Huckman, Robert S., and Eli Strick. "GlaxoSmithKline: Reorganizing Drug Discovery (A)." Harvard Business School Case 605-074. May 2005. https://www.hbs.edu/faculty/Pages/item.aspx?num=32341.

Huggett, Brady. "Reinventing Tech Transfer: US University Technology Transfer Offices Are Adopting New Models in Search of Increased Return on Research Investment." *Nature Biotechnology* 32, no. 12 (December 2014): 1184–91.

Huggett, Brady. "Top US Universities, Institutes for Life Sciences in 2015." *Nature Biotechnology* 35, no. 3 (March 2017): 203.

Hughes, Sally Smith. *Genentech: The Beginnings of Biotech*. Chicago: University of Chicago Press, 2011.

Humphreys, Andrew. "23rd Annual Report: Top 100 Medicines." *Med Ad News*, July 2013. https://www.pharmalive.com/wp-content/uploads/2015/02/144812-MedAdNews-July.pdf.

Iansiti, Marco, and Alan MacCormack. "Developing Products on Internet Time." *Harvard Business Review* 75, no. 5 (September 1997): 108–17.

Imperial College London. *A Founder's Guide to Spinouts—your Guide to Starting a Spinout Company at Imperial College London*. London: Imperial Innovations, 2017. https://www.imperial.tech/media/uploads/files/A_Founders_Guide_to_Spinouts_-_Second_Edition_WEB.pdf.

Instinctif Partners. "Imperial Innovations Group Plc: Year-End Results." Client News. Last modified September 13, 2016. http://lifesciences.instinctif.com/news/2016/10/imperial-innovations-group-plc-year-end-results.

Intel. "Moore's Law and Intel Innovation." Accessed September 17, 2020. https://www.intel.com/content/www/us/en/history/museum-gordon-moore-law.html?wapkw=moore%27s%20law.

Intel Newsroom. "Intel Celebrates 40 Years of Digital Revolution." Last modified November 15, 2011. https://newsroom.intel.com/news-releases/intel-celebrates-40-years-of-digital-revolution/#gs.gfpc6m.

IQVIA Institute for Human Data Science. *The Changing Landscape of Research and Development: Innovation, Drivers of Change, and Evolution of Clinical Trial Productivity*. Parsippany, NJ: IQVIA, 2019. https://www.iqvia.com/insights/the-iqvia-institute/reports/the-changing-landscape-of-research-and-development.

IQVIA Institute for Human Data Science. *The Global Use of Medicine in 2019 and Outlook to 2023: Forecasts and Areas to Watch*. Parsippany, NJ: IQVIA, 2019. https://www.iqvia.com/insights/the-iqvia-institute/reports/the-global-use-of-medicine-in-2019-and-outlook-to-2023.

Jaruzelski, Barry, Robert Chwalik, and Brad Goehle. "The Global Innovation 1000: What the Top Innovators Get Right." *Strategy+Business*, October 30, 2018. https://www.strategy-business.com/media/file/sb93-What-the-Top-Innovators-Get-Right.pdf.

Jones, Catrina M. "Managing Pharmaceutical Research and Development Portfolios: An Empirical Inquiry into Managerial Decision Making in the Context of a Merger." EDB diss., Georgia State University, 2016.

Jones, Mark. "Berg, Boyer, Cohen: The Invention of Recombinant DNA Technology." *Medium*, November 11, 2015. https://medium.com/lsf-magazine/the-invention-of-recombinant-dna-technology-e040a8a1fa22.

Kahneman, Daniel. *Thinking, Fast and Slow*. New York: Penguin Books, 2012.

Kaplan, Steven N., and Per Strömberg. "Venture Capitals as Principals: Contracting, Screening, and Monitoring." *American Economic Review* 91, no. 2 (May 2001): 426–30.

Kaplan, Steven N., and Per Strömberg. "Characteristics, Contracts, and Actions: Evidence from Venture Capitalist Analyses." *Journal of Finance* 59, no. 5 (October 2004): 2177–210.

Kaplan, Steven N., and Per Strömberg. "Financial Contracting Theory Meets the Real World: An Empirical Analysis of Venture Capital Contracts." *Review of Economic Studies* 70, no. 2 (April 2003): 281–315.

Keegan, Karl D. *Biotechnology Valuation: An Introductory Guide*. Hoboken, NJ: John Wiley & Sons, 2008.

Keller Center at Princeton University. "Ten Rules of Innovating at Amazon." Stories. Last modified January 18, 2018. https://kellercenter.princeton.edu/stories/ten-rules-innovating-amazon.

Kennedy, Donald. "Not Wicked, Perhaps, but Tacky." *Science* 297, no. 5585 (August 2002): 1237.

Kerr, Sari Pekkala, William R. Kerr, and Tina Xu. "Personality Traits of Entrepreneurs: A Review of Recent Literature." *Foundations and Trends in Entrepreneurship* 14, no. 3 (July 2018): 279–356.

Kihlstrom, Richard E., and Jean-Jacques Laffont. "A General Equilibrium Entrepreneurial Theory of Firm Formation Based on Risk Aversion." *Journal of Political Economy* 87, no. 4 (August 1979): 719–48.

Kirsch, Donald R. "Therapeutic Drug Development and Human Clinical Trials." In *Biotechnology Entrepreneurship: Leading, Managing and Commercializing Innovative Technologies*, 2nd ed., edited by Craig Shimasaki, 315–50. London: Academic Press, 2020.

Klepper, Steven, and Kenneth L. Simons. "Innovation and Industry Shakeouts." *Business and Economic History* 25, no. 1 (Fall 1996): 81–89.

Knapp, Jake, John Zeratsky, and Braden Kowitz. *Sprint: How to Solve Big Problems and Test New Ideas in Just Five Days*. New York: Simon & Schuster, 2016.

Kneller, Robert. "The Importance of New Companies for Drug Discovery: Origins of a Decade of New Drugs." *Nature Reviews Drug Discovery* 9, no. 11 (November 2010): 867–82.

Knight, Frank H. *Risk, Uncertainty, and Profit*. Boston: Houghton Mifflin, 1921.

Knight, Jonathan. "Biotech Woes Set to Hit Academics." *Nature* 418, no. 6893 (July 2002): 5.

Kornberg, Arthur. *The Golden Helix: Inside Biotech Ventures*. Sausalito, CA: University of Science Books, 1995.

Kornish, Laura J., and Karl T. Urlich. "The Importance of the Raw Idea in Innovation: Testing the Sow's Ear Hypothesis." *Journal of Marketing Research* 51, no. 1 (February 2014): 14–26.

Kortum, Samuel, and Josh Lerner. "Assessing the Contribution of Venture Capital to Innovation." *RAND Journal of Economics* 31, no. 4 (Winter 2000): 674–92.

KPMG International Cooperative. *Site Selection for Life Sciences Companies in Europe, 2018 Edition*. Amstelveen, Netherlands: KPMG AG, 2018. https://assets.kpmg/content/dam/kpmg/ch/pdf/site-selection-for-life-sciences-companies-europe-2018-en.pdf.

Kumar, Harpal. "Years of Crucial Investment Could Fall by the Wayside." *Times* (London), September 8, 2010, 16.

Lawrence, Stacy. "R&D Growth Stalls." *Acumen Journal of Sciences* 1, no. 2 (August/September 2003): 22.

Lawrence, Stacy, and Riku Lahteenmaki. "Public Biotech 2013—the Numbers." *Nature Biotechnology* 32, no. 7 (July 2014): 626–32.

Lazard. "Lazard Releases Global Healthcare Leaders Study." News release, May 15, 2017. https://www.lazard.com/media/450175/lazard-releases-global-healthcare-leaders-study-51517.pdf.

Lazear, Edward P. "Balanced Skills and Entrepreneurship." *American Economic Review* 94, no. 2 (May 2004): 208–11.

Lazear, Edward P. "Entrepreneurship." *Journal of Labor Economics* 23, no. 4 (October 2005): 649–80.

Leaf, Clifton. "Why We're Losing the War on Cancer (and How to Win It)." *Fortune*, March 22, 2004. https://fortune.com/2004/03/22/cancer-medicines-drugs-health/.

Leber, Jessica. "How Google's Moonshot X Division Helps Its Employees Embrace Failure." *Fast Company*, April 14, 2016. https://www.fastcompany.com/3058866/how-googles-moonshot-x-division-helps-its-employees-embrace-failure.

Ledford, Heidi. "Virtual Reality." *Nature* 498, no. 7452 (June 2013): 127–29.

Lenzen, Manfred, et al. "Global Socio-Economic Losses and Environmental Gains from the Coronavirus Pandemic." *PLOS ONE* 15, no. 7 (July 2020): 1–13.

Leonard-Barton, Dorothy. "Core Capabilities and Core Rigidities: A Paradox in Managing New Product Development." In "Strategy Process: Managing Corporate Self-Renewal," edited by Dan Schendel and Derek Channon, special issue. *Strategic Management Journal* 13, S1 (Summer 1992): 111–25.

Lerner, Josh. "Venture Capitalists and the Oversight of Private Firms." *Journal of Finance* 50, no. 1 (March 1995): 301–18.

Lerner, Josh, and Robert P. Merges. "The Control of Technology Alliances: An Empirical Analysis of the Biotechnology Industry." *Journal of Industrial Economics* 46, no. 2 (March 27, 2003): 125–56.

Lerner, Josh, Hilary Shane, and Alexander Tsai. "Do Equity Financing Cycles Matter? Evidence from Biotechnology Alliances." *Journal of Financial Economics* 67, no. 3 (March 2003): 411–46.

Li, Jie Jack. *Top Drugs: Their History, Pharmacology, and Syntheses*. New York: Oxford University Press, 2015.

Lipman, Arthur G. Review of *Drug Repurposing and Repositioning: Workshop Summary*, by Sarah H. Beachy, Samuel G. Johnson, Steven Olson, and Adam C. Berger. *Journal of Pain & Palliative Care Pharmacotherapy* 29, no. 1 (February 2015): 81.

Liu, Angus. "Takeda, Schrödinger Form Multiprogram Drug Discovery Pact." *Fierce Biotech*, July 20, 2017. https://www.fiercebiotech.com/cro/takeda-schrodinger-form-multiprogram-drug-discovery-pact.

Loch, Christoph H., Arnoud De Meyer, and Michael T. Pich. *Managing the Unknown: A New Approach to Managing High Uncertainty and Risks in Projects*. Hoboken, NJ: John Wiley & Sons, 2006.

Lonberg, Nils, and Dennis Huszar. "Human Antibodies from Transgenic Mice." *International Reviews of Immunology* 13, no. 1 (1995): 65–93.

Lonberg, Nils, Lisa D. Taylor, Fiona A. Harding, Mary Trounstine, Kay M. Higgins, Stephen R. Schramm, Chiung-Chi Kuo, et al. "Antigen-Specific Human Antibodies from Mice Comprising Four Distinct Genetic Modifications." *Nature* 368, no. 6474 (April 1994): 856–59.

Longaker, Michael T., Laurence C. Baker, and Henry T. Greely. "Proposition 71 and CIRM—Assessing the Return on Investment." *Nature Biotechnology* 25, no. 5 (May 2007): 513–21.

Longman, Roger. "Guest Commentary: Cost Steamroller." *BioCentury*, June 9, 2014. https://www.biocentury.com/article/248976.

Looney, William. "Pharm Exec's Top 50 Companies 2016." *Pharm Exec*, July 26, 2016. https://www.pharmexec.com/view/2016-pharm-exec-50.

Lovallo, Dan, and Daniel Kahneman. "Delusions of Success: How Optimism Undermines Executives' Decisions." *Harvard Business Review* 81, no. 7 (July 2003): 56–63.

Lowry, Michelle. "Why Does IPO Volume Fluctuate So Much?" *Journal of Financial Economics* 67, no. 1 (January 2003): 3–40.

Löwy, Ilana. *Between Bench and Bedside: Science, Healing and Inteleukin-2 in a Cancer Ward.* Cambridge, MA: Harvard University Press, 1996.

Lu, Ruei-Min et al. "Development of Therapeutic Antibodies for the Treatment of Diseases." *Journal of Biomedical Science* 27, no. 1 (2020): 1–30.

Luessen, Henrik, ed. *Starting a Business in the Life Sciences: From Idea to Market.* Aulendorf, Germany: Editio Cantor Verlag, 2003.

Ma, Dan. "When Bill Gates Walks into a Bar." *Introductory Statistics,* September 4, 2011. https:// introductorystats.wordpress.com/2011/09/04/when-bill-gates-walks-into-a-bar/.

Manso, Gustavo. "Creating Incentives for Innovation." *California Management Review* 60, no. 1 (November 2017): 18–32.

Markou, Panos, Stylianous Kavadias, and Nektarios Oraiopoulos. "Rival Signals and Project Selection: Insights from the Drug Development Process." *SSRN* (June 29, 2020): 1–67, https://ssrn.com/abstract=3225056.

Marks, Lara, ed. *Engineering Health: How Biotechnology Changed Medicine.* London: Royal Society of Chemistry, 2017.

Marks, Lara V. *The Lock and Key of Medicine: Monoclonal Antibodies and the Transformation of Healthcare.* New Haven, CT: Yale University Press, 2015.

Mason, Richard, and Donald L. Drakeman. "Comment on 'Fishing for Sharks: Partner Selection in Biopharmaceutical R&D Alliances' by Diestre and Rajagopalan." *Strategic Management Journal* 35, no. 10 (October 2014): 1564–65.

Mason, Richard, Nicos Savva, and Stefan Scholtes. "The Economics of Licensing Contracts." *Nature Biotechnology* 26, no. 8 (August 2008): 855–57.

Maula, Markku, and Gordon Murray. "Corporate Venture Capital and the Creation of US Public Companies: The Impact of Sources of Venture Capital on the Performance of Portfolio Companies." In *Creating Value: Winners in the New Business Environment,* edited by Michael A. Hitt, Raphael Amit, Charles E. Lucier, and Robert D. Nixon, 164–87. Oxford: Blackwell Publishers, 2002.

Maycotte, H. O. "Education vs. Entrepreneurship: Which Path Wins?" *Forbes,* June 2, 2015. https://www.forbes.com/sites/homaycotte/2015/06/02/education-vs-entrepreneurship-which-path-wins/#2c3660714cdc.

McCallister, Erin. "Independents' Day: After Their First Solo Launch, Biotechs Look to New Partnering Strategies." *BioCentury,* May 3, 2019. https://www.biocentury.com/article/302 071/how-the-next-wave-of-independent-companies-are-adapting-for-long-term-growth.

McGrath, Rita Gunther. "Exploratory Learning, Innovative Capacity, and Managerial Oversight." *Academy of Management Journal* 44, no. 1 (February 2001): 118–31.

McGrath, Rita Gunther, and Atul Nerkar. "Real Options Reasoning and a New Look at the R&D Investment Strategies of Pharmaceutical Firms." *Strategic Management Journal* 25, no. 1 (January 2004): 1–21.

McKelvey, Maureen D. *Evolutionary Innovation: The Business of Biotechnology.* Oxford: Oxford University Press, 1996.

McKinsey-Lehman Brothers Report. *The Fruits of Genomics.* New York: Lehman Brothers, 2000.

Medical Research Council. *MRC Delivery Plan: 2011/12 to 2014/15.* November 2010. https:// mrc.ukri.org/publications/browse/delivery-plan-201112-201415/.

Mittra, James. "Impact of the Life Sciences on Organisation and Management of R&D in Large Pharmaceutical Firms." *International Journal of Biotechnology* 10, no. 5 (November 2008): 416–40.

Mobius Life Sciences. *Realignment: UK Life Science Start-Up Report 2012.* Nottingham, UK: Mobius Life Sciences, 2012. https://biocity.co.uk/wp-content/uploads/pdf/realignm ent-uk-life-science-start-up-report-2012.pdf.

Molineux, Graham, MaryAnn Foote, and Tara Arvedson, eds. *Twenty Years of G-CSF: Clinical and Nonclinical Discoveries.* Basel, Switzerland: Springer Basel, 2012.

Morrison, Chris. "2019 Biotech IPOs: Party On." *Nature Reviews Drug Discovery* 19, no. 1 (January 2020): 6–9.

Morrison, Chris, and Riku Lähteenmäki. "Public Biotech 2018—The Numbers." *Nature Biotechnology* 37, no. 7 (July 2019): 714–21.

Mowery, David C., Richard R. Nelson, Bhaven N. Sampat, and Arvids A. Ziedonis. *Ivory Tower and Industrial Innovation: University-Industry Technology Transfer before and after the Bayh-Dole Act.* Stanford, CA: Stanford University Press, 2015.

Munos, Bernard. "Lessons from 60 Years of Pharmaceutical Innovation." *Nature Reviews Drug Discovery* 8, no. 12 (December 2009): 959–68.

Murphy, Kevin M., and Robert H. Topel, eds. *Measuring the Gains from Medical Research: An Economic Approach.* Chicago: University of Chicago Press, 2003.

Murphy, Kevin M., and Robert H. Topel. "The Economic Value of Medical Research." In Murphy and Topel, *Measuring the Gains from Medical Research*, 41–73, 2003.

Myriad Genetics. "Myriad Genetics Selects Lundbeck as European Partner for Flurizan." News release, May 22, 2008. https://investor.myriad.com/static-files/e02cacd4-b32a-486a-9e46-338126339823.

Nanda, Ramana, and Matthew Rhodes-Kropf. "Innovation Policies." In *Entrepreneurship, Innovation, and Platforms.* Vol. 37 of *Advances in Strategic Management*, edited by Jeffrey Furman, Annabelle Gawer, Brian S. Silverman, and Scott Stern, 37–80. Somerville, MA: Emerald Publishing, 2017.

Nanda, Ramana, and Matthew Rhodes-Kropf. "Investment Cycles and Startup Innovation." *Journal of Financial Economics* 110, no. 2 (November 2013): 403–18.

National Academies of Sciences, Engineering, and Medicine. *Deriving Drug Discovery Value from Large-Scale Genetic Bioresources: Proceedings of a Workshop.* Washington, DC: National Academies Press, 2016.

National Center for Advancing Translational Sciences. *Discovering New Therapeutic Uses for Existing Molecules.* Bethesda, MD: National Institutes of Health, 2015. https://ncats.nih.gov/files/NTU-factsheet.pdf.

National Center for Advancing Translational Sciences. "NCATS Announces Funding Opportunities to Repurpose Existing Drugs through Public-Private Partnerships." Last modified February 2017. https://ncats.nih.gov/news/releases/2017/drug-repurpose-partn erships.

National Institute for Health and Care Excellence. "NICE Publishes Updated Principles." January 30, 2020. https://www.nice.org.uk/news/article/nice-publishes-updated-principles.

National Institute for Health and Care Excellence. "Our Charter." Accessed September 17, 2020. https://www.nice.org.uk/about/who-we-are/our-charter.

National Institute for Health and Care Excellence. 2020. "Our Principles." https://www.nice.org.uk/about/who-we-are/our-principles.

National Institute for Health and Clinical Excellence. *Social Value Judgements: Principles for the Development of NICE Guidance.* 2nd ed. London: NICE, 2005. https://www.nice.org.uk/Media/Default/about/what-we-do/research-and-development/Social-Value-Judgements-principles-for-the-development-of-NICE-guidance.docx.

National Institutes of Health. *Estimates of Funding for Various Research, Condition, and Disease Categories (RCDC).* March 7, 2014. https://report.nih.gov/categorical_spending.aspx.

National Institutes of Health. "Our Society." Last modified May 1, 2018. https://www.nih.gov/about-nih/what-we-do/impact-nih-research/our-society.

National Research Council. *Managing University Intellectual Property in the Public Interest.* Washington, DC: National Academies Press, 2011.

Naylor, Stephen, and Kirkwood A. Pritchard Jr. "The Reality of Virtual Pharmaceutical Companies." *Drug Discovery World* (Summer 2019): 8–14. https://www.ddw-online.com/business/p323009-the-reality-of-virtual-pharmaceutical-companies.html.

Nelsen, Lita. "Technology Licensing Office." Reports to the President, March 18, 2015. http://web.mit.edu/annualreports/pres05/03.18.pdf.

Nelson, Richard R. "Uncertainty, Learning, and the Economics of Parallel Research and Development Efforts." *Review of Economics and Statistics* 43, no. 4 (November 1961): 351–64.

Nelson, Richard R. "The Market Economy, and the Scientific Commons." *Research Policy* 33, no. 3 (April 2004): 455–71.

Neuberger, Arthur, Nektarios Oraiopoulos, and Donald L. Drakeman. "Lemons, or Squeezed for Resources? Information Symmetry and Asymmetric Resources in Biotechnology." *Frontiers in Pharmacology* 8, no. 338 (June 2017): 1–4.

Neuberger, Arthur, Nektarios Oraiopoulos, and Donald L. Drakeman. "Renovation as Innovation: Is Repurposing the Future of Drug Discovery Research?" *Drug Discovery Today* 24, no. 1 (January 2019): 1–3.

New Jersey's Science & Technology University. *Program Evaluation: New Jersey Technology Business Tax Certificate Transfer Program.* Trenton: New Jersey Economic Development Authority, 2010. https://www.nj.gov/transparency/reports/pdf/NJ%20Technology%20Business%20Tax%20Certificate%20Transfer%20Program%5B1%5D.pdf.

Nicholson, Sean. "Financing Research and Development." In *The Oxford Handbook of the Economics of the Biopharmaceutical Industry*, edited by Patricia M. Danzon and Sean Nicholson, 47–74. New York: Oxford University Press, 2012.

Nicholson, Sean, Patricia M. Danzon, and Jeffrey McCullough. "Biotech-Pharmaceutical Alliances as a Signal of Asset and Firm Quality." *Journal of Business* 78, no. 4 (July 2005): 1433–64.

"The Nobel Prize in Physiology or Medicine 2018." August 2018. Nobel Media AB. https://www.nobelprize.org/prizes/medicine/2018/summary/.

"Nobel Work That Galvanized an Industry." *Nature Biotechnology* 36, no. 11 (2018): 1023.

Nosengo, Nicola. "Can You Teach Old Drugs New Tricks?" *Nature* 534, no. 7607 (June 2016): 314–16.

Obschonka, Martin, Håkan Andersson, Rainer K. Silbereisen, and Magnus Sverke. "Rule-Breaking, Crime, and Entrepreneurship: A Replication and Extension Study with 37-Year Longitudinal Data." *Journal of Vocational Behavior* 83, no. 3 (December 2013): 386–96.

Office of the New York State Comptroller. *The Economic Impact of the Biotechnology and Pharmaceutical Industries in New York.* New York, 2005. https://web.osc.state.ny.us/osdc/biotechreport.pdf.

Olson, Maynard V. "The Human Genome Project: A Player's Perspective." *Journal of Molecular Biology* 319, no. 4 (June 2002): 931–42.

Oraiopoulos, Nektarios, and William C. N. Dunlop. "When Science Is Not Enough: A Framework Towards More Customer-Focused Drug Development." *Advances in Therapy* 34 (2017): 1572–83.

Oraiopoulos, Nektarios, and Stylianos Kavadias. "Is Diversity (Un)Biased? Project Selection Decisions in Executive Committees." *Manufacturing & Service Operations Management* 22, no. 5 (2020): 906–24.

Organisation for Economic Co-operation and Development. *Key Biotechnology Indicators.* December 2011. http://www.oecd.org/science/inno/49303992.pdf.

Organisation for Economic Co-operation and Development. "Number of Firms Active in Biotechnology, 2006–2018." Last modified October 2020. https://www.oecd.org/sti/KBI1-number-of-firms-active-in-biotech-2020.xlsx.

Organisation for Economic Co-operation and Development. "Pharmaceutical Spending (Indicator)." Accessed September 29, 2020. https://data.oecd.org/healthres/pharmaceutical-spending.htm.

Owen, Geoffrey, and Michael M. Hopkins. *Science, the State, & the City: Britain's Struggle to Succeed in Biotechnology.* Oxford: Oxford University Press, 2016.

Ozmel, Umit, and Isin Guler. "Small Fish, Big Fish: The Performance Effects of the Relative Standing in Partners' Affiliate Portfolios." *Strategic Management Journal* 36, no. 13 (December 2015): 2039–57.

Ozmel, Umit, Jeffrey J. Reuer, and Ranjay Gulati. "Signals across Multiple Networks: How Venture Capital and Alliance Networks Affect Interorganizational Collaboration." *Academy of Management Journal* 56, no. 3 (June 2013): 852–66.

Pammolli, Fabio, Laura Magazzini, and Massimo Riccaboni. "The Productivity Crisis in Pharmaceutical R&D." *Nature Reviews Drug Discovery* 10, no. 6 (June 2011): 428–38.

Papadopoulos, Stelios. "Evolving Paradigms in Biotech IPO Valuations." *Nature Biotechnology* 19, no. 6 (June 2001): BE 18–19.

Paul, Steven M., Daniel S. Mytelka, Christopher T. Dunwiddie, Charles C. Persinger, Bernard H. Munos, Stacy R. Lindborg, and Aaron L. Schacht. "How to Improve R&D Productivity: The Pharmaceutical Industry's Grand Challenge." *Nature Reviews Drug Discovery* 9, no. 3 (March 2010): 203–14.

Pharma-Medical Science College of Canada. "Canadian Biotechnology Company List." Last modified January 30, 2020. https://www.pharmamedical.ca/canadian-biotechnology-company-list/.

Philippidis, Alex. "Top 10 Molecular Millionaires." *GEN: Genetic Engineering & Biotechnology News*, January 2, 2017. http://www.genengnews.com/insight-and-intelligenceand153/top-10-molecular-millionaires/77899901.

Philippidis, Alex. "The Top 15 Best-Selling Drugs of 2016." *GEN: Genetic Engineering & Biotechnology News*, March 6, 2017. https://www.genengnews.com/the-lists/the-top-15-best-selling-drugs-of-2016/77900868.

Philippidis, Alex. "The Top 15 Best-Selling Drugs of 2017." *GEN: Genetic Engineering & Biotechnology News*, March 12, 2018. https://www.genengnews.com/a-lists/the-top-15-best-selling-drugs-of-2017/.

Pich, Michael T., Christoph H. Loch, and Arnoud De Meyer. "On Uncertainty, Ambiguity, and Complexity in Project Management." *Management Science* 48, no. 88 (August 2002): 1008–23.

Pisano, Gary P. "The R&D Boundaries of the Firm: An Empirical Analysis." *Administrative Science Quarterly* 35, no. 1 (March 1990): 153–76.

Pisano, Gary P. "R&D Performance, Collaborative Arrangements and the Market for Know-How: A Test of the 'Lemons' Hypothesis in Biotechnology." Working paper, Harvard Business School, Cambridge, MA, July 1997.

Pisano, Gary P. "Can Science Be a Business? Lessons from Biotech." *Harvard Business Review* 84, no. 10 (October 2006): 114–24.

Pisano, Gary P. *Science Business: The Promise, the Reality, and the Future of Biotech.* Brighton, MA: Harvard Business School Press, 2006.

Pisano, Gary P., and Roberto Verganti. "Which Kind of Collaboration Is Right for You?" *Harvard Business Review* 86, no. 12 (December 2008): 78–86.

Pitts, Peter J. "A Flawed Study Depicts Drug Companies as Profiteers." *Wall Street Journal*, October 9, 2017. https://www.wsj.com/articles/a-flawed-study-depicts-drug-companies-as-profiteers-1507590936.

Pollack, Andrew. "Three Universities Join Researcher to Develop Drugs." *New York Times*, July 31, 2003, C1.

Pollack, Andrew. "Myriad Genetics Stops Work on Alzheimer's Drug." *New York Times*, July 1, 2008. https://www.nytimes.com/2008/07/01/business/01gene.html.

Powell, Walter W., Kenneth W. Koput, and Laurel Smith-Doerr. "Interorganizational Collaboration and the Locus of Innovation: Networks of Learning in Biotechnology." *Administrative Science Quarterly* 41, no. 1 (March 1996): 116–45.

Power, Aidan, Adam C. Berger, and Geoffrey S. Ginsburg. "Genomics-Enabled Drug Repositioning and Repurposing: Insights from an IOM Roundtable Activity." *JAMA* 311, no. 20 (May 2014): 2063–64.

Prasad, Vinay, and Sham Mailankody. "Research and Development Spending to Bring a Single Cancer Drug to Market and Revenues after Approval." *JAMA Internal Medicine* 177, no. 11 (November 2017): 1569–75.

Preqin. *The Venture Capital Industry: A Preqin Special Report, October 2010*. London: Preqin, 2010. https://docs.preqin.com/reports/Preqin_Private_Equity_Venture_Report_Oct2010.pdf.

Pressman, Loris, David Roessner, Jennifer Bond, Sumiye Okubo, and Mark Planting. *The Economic Contribution of University/Nonprofit Inventions in the United States: 1996–2013*. Washington, DC: Biotechnology Industry Organization, 2015. https://www.bio.org/sites/default/files/legacy/bioorg/docs/BIO_2015_Update_of_I-O_Eco_Imp.pdf.

Princeton University. "Middlekauff Named Executive in Residence (XIR) to Support Tech-Transfer Experience." Last modified October 8, 2014. https://research.princeton.edu/news/middlekauff-named-executive-residence-xir-support-tech-transfer-experience.

Qian, Yingyi, and Chenggang Xu. "Innovation and Bureaucracy under Soft and Hard Budget Constraints." *Review of Economic Studies* 65, no. 1 (January 1998): 151–64.

Rai, Arti K., and Rebecca S. Eisenburg. "Bayh-Dole Reform and the Progress of Biomedicine." *American Scientist* 91, no. 1 (January/February 2003): 52–59.

Rasmussen, Nicolas. *Gene Jockeys: Life Science and the Rise of Biotech Enterprise*. Baltimore, MD: Johns Hopkins University Press, 2014.

Redelmeier, Donald A., and Eldar Shafir. "Medical Decision Making in Situations That Offer Multiple Alternatives." *JAMA* 273, no. 4 (January 1995): 302–5.

Reichert, Janice M. "Monoclonal Antibodies as Innovative Therapeutics." *Current Pharmaceutical Biotechnology* 9, no. 6 (2008): 423–30.

Reiss, Thomas. "Drug Discovery of the Future: The Implications of the Human Genome Project." *Trends in Biotechnology* 19, no. 12 (December 2001): 496–99.

Remington, Michael J. "A Board's Primer on Technology Transfer." *Trusteeship* 10, no. 6 (November/December 2002): 13–18.

Reslinski, Michael A., and Bernhard S. Wu. "The Value of Royalty." *Nature Biotechnology* 34, no. 7 (July 2016): 685–90.

Reuer, Jeffrey J., and Roberto Ragozzino. "The Choice between Joint Ventures and Acquisitions: Insights from Signaling Theory." *Organization Science* 23, no. 4 (July/August 2012): 1175–90.

Reuer, Jeffrey J., and Roberto Ragozzino. "Signals and International Alliance Formation: The Roles of Affiliations and International Activities." *Journal of International Business Studies* 45, no. 3 (April 2014): 321–37.

Rhodes, Rosamond, Margaret Battin, and Anita Silvers, eds. *Medicine and Social Justice: Essays on the Distribution of Health Care*. New York: Oxford University Press, 2002.

Richards, Graham. *Spin-Outs: Creating Businesses from University Intellectual Property*. Hampshire, UK: Harriman House, 2009.

Ridley, Matt. *Genome: The Autobiography of a Species in 23 Chapters*. New York: HarperCollins, 2000.

Robbins-Roth, Cynthia. *From Alchemy to IPO: The Business of Biotechnology*. Cambridge, MA: Perseus Publishing, 2000.

Robinson, David T., and Toby E. Stuart. "Financial Contracting in Biotech Strategic Alliances." *Journal of Law and Economics* 50, no. 3 (August 2007): 559–96.

Roijakkers, Nadine, and John Hagedoorn. "Inter-firm R&D Partnering in Pharmaceutical Biotechnology since 1975: Trends, Patterns, and Networks." *Research Policy* 35, no. 3 (April 2006): 431–46.

Rothaermel, Frank T. "Complementary Assets, Strategic Alliances, and the Incumbent's Advantage: An Empirical Study of Industry and Firm Effects in the Biopharmaceutical Industry." *Research Policy* 30, no. 8 (October 2001): 1235–51.

Rothaermel, Frank T., and Warren Boeker. "Old Technology Meets New Technology: Complementarities, Similarities, and Alliance Formation." *Strategic Management Journal* 29, no. 1 (January 2008): 47–77.

Rothaermel, Frank T., and David L. Deeds. "Exploration and Exploitation Alliances in Biotechnology: A System of New Product Development." *Strategic Management Journal* 25, no. 3 (March 2004): 201–21.

Sagonowsky, Eric. "The Decade's Top 10 Patent Losses, Worth a Whopping $915B in Lifetime Sales." *Fierce Pharma*, August 17, 2017. https://www.fiercepharma.com/pharma/decade-s-top-10-patent-losses-featuring-seismic-sales-shifts.

Sagonowsky, Eric. "The Top 10 Most-Expensive Meds in the U.S.—and They're Not the Usual Suspects." *Fierce Pharma*, June 13, 2019. https://www.fiercepharma.com/pharma/most-expensive-meds-u-s-topped-by-novartis-and-spark-gene-therapies.

Sah, Raaj Kumar, and Joseph E. Stiglitz. "The Architecture of Economic Systems: Hierarchies and Polyarchies." *American Economic Review* 76, no. 4 (September 1986): 716–27.

Sahlman, William A. "Aspects of Financial Contracting in Venture Capital." *Journal of Applied Corporate Finance* 1, no. 2 (Summer 1988): 23–36.

Sahlman, William A. "The Structure and Governance of Venture-Capital Organizations." *Journal of Financial Economics* 27, no. 2 (October 1990): 473–521.

Sahlman, William A. "Risk and Reward in Venture Capital." Harvard Business School Background Note 811-036. December 2010. https://www.hbs.edu/faculty/Pages/item.aspx?num=39710.

Sainsbury, Lord David. "A Cultural Change in UK Universities." *Science* 296, no. 5575 (June 2002): 1929.

Samila, Sampsa, and Olav Sorenson. "Venture Capital, Entrepreneurship, and Economic Growth." *Review of Economics and Statistics* 93, no. 1 (February 2011): 338–49.

Samson, Karel J. *Scientists as Entrepreneurs: Organizational Performance in Scientist-Started New Ventures.* Boston: Kluwer Academic Publishers, 1991.

S&P Global Platform. S&P Capital IQ Platform. https://www.spglobal.com/marketintelligence/en/solutions/sp-capital-iq-platform.

Scannell, Jack W., Alex Blanckley, Helen Boldon, and Brian Warrington. "Diagnosing the Decline in Pharmaceutical R&D Efficiency." *Nature Reviews Drug Discovery* 11, no. 3 (March 2012): 191–200.

Schlapp, Jochen, Nektarios Oraiopoulos, and Vincent Mak. "Resource Allocation Decisions under Imperfect Evaluation and Organizational Dynamics." *Management Science* 61, no. 9 (2015): 2139–59.

Schuhmacher, Alexander, Oliver Gassmann, and Markus Hinder. "Changing R&D Models in Research-Based Pharmaceutical Companies." *Journal of Translational Medicine* 14, no. 105 (April 2016): 1–11.

Schwartz, Barry. *The Paradox of Choice: Why More Is Less.* New York: HarperCollins, 2004.

Schweitzer, Stuart O. *Pharmaceutical Economics and Policy.* 3rd ed. New York: Oxford University Press, 2018.

Searle, Susan, Brian Graves, and Chris Towler. "Commercializing Biotechnology in the UK." *Bioentrepreneur*, January 29, 2003. https://www.nature.com/articles/bioent710.pdf.

Seneca, Joseph J., and Will Irving. *The Economic Benefits of the New Jersey Stem Cell Research Initiative*. New Brunswick, NJ: Rutgers, 2005. https://pdfs.semanticscholar.org/5fdf/1098479d326f963d83caee3f30f94bb4b214.pdf.

Seneca, Joseph J., and Will Irving. *Updated Economic Benefits of the New Jersey Stem Cell Capital Projects and Research Bond Acts*. New Brunswick, NJ: Rutgers, 2007. https://bloustein.rutgers.edu/wp-content/uploads/2015/03/stemcelloct07.pdf.

Seru, Amit. "Firm Boundaries Matter: Evidence from Conglomerates and R&D Activity." *Journal of Financial Economics* 111, no. 2 (February 2014): 381–405.

Service, Robert F. "Surviving the Blockbuster Syndrome." *Science* 303, no. 5665 (March 2004): 1796–99.

Sharad, Shashwat, and Suman Kapur, eds. *Antisense Therapy*. London: IntechOpen, 2019. https://www.intechopen.com/books/antisense-therapy/antisense-therapy-an-overview.

Shimasaki, Craig, ed. *Biotechnology Entrepreneurship: Leading, Managing and Commercializing Innovative Technologies*. 2nd ed. London: Academic Press, 2020.

Siegel, Donald S., Reinhilde Veugelers, and Mike Wright. "Technology Transfer Offices and Commercialization of University Intellectual Property: Performance and Policy Implications." *Oxford Review of Economic Policy* 23, no. 4 (Winter 2007): 640–60.

Siler, Kyle, Kirby Lee, and Lisa Bero. "Measuring the Effectiveness of Scientific Gatekeeping." *Proceedings of the National Academy of Sciences* 112, no. 2 (January 2015): 360–65.

Silva, Olmo. "The Jack-of-All-Trades Entrepreneur: Innate Talent or Acquired Skill?" *Economics Letters* 97, no. 2 (November 2007): 118–23.

Simon, Françoise, and Glen Giovannetti. *Managing Biotechnology: From Science to Market in the Digital Age*. Hoboken, NJ: John Wiley & Sons, 2017.

"Sir Gregory P. Winter: Nobel Lecture." December 8, 2018. Nobel Media AB. https://www.nobelprize.org/prizes/chemistry/2018/winter/lecture/.

Skloot, Rebecca. *The Immortal Life of Henrietta Lacks*. New York: Crown Publishing, 2010.

Sommer, Svenja C., and Christoph H. Loch. "Selectionism and Learning in Projects with Complexity and Unforeseeable Uncertainty." *Management Science* 50, no. 10 (October 2004): 1334–47.

Stanford University. "9.1 Inventions, Patents, and Licensing." Research Policy Handbook. Last modified June 19, 2013. https://doresearch.stanford.edu/policies/research-policy-handbook/intellectual-property/inventions-patents-and-licensing.

Stasior, Jonathan, Brian Machinist, and Michael Esposito. *Valuing Pharmaceutical Assets: When to Use NPV vs rNPV*. London: Alacrita Consulting, 2018. https://www.alacrita.com/whitepapers/valuing-pharmaceutical-assets-when-to-use-npv-vs-rnpv.

Station, Tracy. "What Keeps 88% of Biopharma Executives Up at Night? One Guess, and It's Pricing." *Fierce Pharma*, May 18, 2017. https://www.fiercepharma.com/pharma/what-keeps-88-biopharma-execs-up-at-night-one-guess-and-it-s-pricing.

Staw, Barry M. "Knee-Deep in the Big Muddy: A Study of Escalating Commitment to a Chosen Course of Action." *Organizational Behavior and Human Performance* 16, no. 1 (June 1976): 27–44.

Stein, Rob. "At $2.1 Million, New Gene Therapy Is the Most Expensive Drug Ever." *Shots: Health News from NPR*, May 24, 2019. https://www.npr.org/sections/health-shots/2019/05/24/725404168/at-2-125-million-new-gene-therapy-is-the-most-expensive-drug-ever.

Stevens, Ashley J., Jonathan J. Jensen, Katrine Wyller, Patrick C. Kilgore, Sabarni Chatterjee, and Mark L. Rohrbaugh. "The Role of Public-Sector Research in the Discovery of Drugs and Vaccines." *The New England Journal of Medicine* 364, no. 6 (February 2011): 535–41.

Stewart, Wayne H., Jr., and Philip L. Roth. "A Meta-analysis of Achievement Motivation Differences between Entrepreneurs and Managers." *Journal of Small Business Management* 45, no. 4 (October 2007): 401–21.

Strittmatter, Stephen M. "Overcoming Drug Development Bottlenecks with Repurposing: Old Drugs Learn New Tricks." *Nature Medicine* 20, no. 6 (June 2014): 590–91.

Stuart, Toby E., Ha Hoang, and Ralph C. Hybels. "Interorganizational Endorsements and the Performance of Entrepreneurial Ventures." *Administrative Science Quarterly* 44, no. 2 (June 1999): 315–49.

Stuart, Toby E., Salih Zeki Ozdemir, and Waverly W. Ding. "Vertical Alliance Networks: The Case of University-Biotechnology-Pharmaceutical Alliance Chains." *Research Policy* 36, no. 4 (May 2007): 477–98.

Sulston, John, and Georgina Ferry. *The Common Thread: A Story of Science, Politics, Ethics and the Human Genome.* Washington, DC: Joseph Henry Press, 2002.

Teitelman, Robert. *Gene Dreams: Wall Street, Academia, and the Rise of Biotechnology.* New York: Basic Books, 1989.

Thaler, Richard H. *Misbehaving: The Making of Behavioral Economics.* New York: W. W. Norton, 2015.

Thibodeaux, Clare. "New Repurposing Research Funding Opportunity for Rare Diseases." Cures within Reach. News/Events. Last modified January 11, 2016. https://web.archive.org/web/20160518151029/http://cureswithinreach.org/newsroom/blog-re-rx/460-new-repurposing-research-funding-opportunity-for-rare-diseases.

Thiel Fellowship. Accessed September 14, 2020. http://thielfellowship.org/.

Thomas, David, and Chad Wessel. *Emerging Therapeutic Company Investment and Deal Trends.* Washington, DC: Biotechnology Industry Organization, 2016. http://go.bio.org/rs/490-EHZ-999/images/BIO_Emerging_Therapeutic_Company_Report_2006_2015_Final.pdf.

Thursby, Jerry G., and Marie C. Thursby. "Who Is Selling the Ivory Tower? Sources of Growth in University Licensing." *Management Science* 48, no. 1 (January 2002): 90–104.

Toole, Andrew A. "Does Public Scientific Research Complement Private Investment in Research and Development in the Pharmaceutical Industry?" *Journal of Law and Economics* 50, no. 1 (February 2007): 81–104.

Troy, Lisa C., Tanawat Hirunyawipada, and Audesh K. Paswan. "Cross-Functional Integration and New Product Success: An Empirical Investigation of the Findings." *Journal of Marketing* 72, no. 6 (November 2008): 132–46.

Tushman, Michael L., and Philip Anderson. "Technological Discontinuities and Organizational Environments." *Administrative Science Quarterly* 31, no. 3 (September 1986): 439–65.

Tuzman, Karen Tkach. "Rise of the First-Time Biotech CEO." *BioCentury*, December 19, 2019. https://www.biocentury.com/article/304098/the-management-crunch-in-biotech-is-creating-a-heyday-for-first-time-ceos.

UK Research and Innovation. "Biomedical Catalyst: Developmental Pathway Funding Scheme (DPFS) Outline: Mar 2021." Last modified March 25, 2021. https://mrc.ukri.org/funding/browse/biomedical-catalyst-dpfs/biomedical-catalyst-developmental-pathway-funding-scheme-dpfs-outline-mar-2021/.

University of Cambridge. *Annual Review 2017.* Cambridge: Cambridge Enterprise, 2017. https://www.enterprise.cam.ac.uk/wp-content/uploads/2015/04/updated-Annual-Report-website.pdf.

Urquhart, Lisa. "Top Companies and Drugs by Sales in 2019." *Nature Reviews Drug Discovery* 19, no. 4 (April 2020): 228.

U.S. Department of Commerce. *A Survey of the Use of Biotechnology in U.S. Industry.* November 2003. https://www.bis.doc.gov/index.php/documents/technology-evaluation/25-a-survey-of-the-use-of-biotechnology-in-u-s-industry-2003/file.

U.S. Department of Health and Human Services, National Institutes of Health. *National Institutes of Health Response to the Conference Report Request for a Plan to Ensure Taxpayers' Interests Are Protected.* Washington, DC, 2001.

U.S. Food & Drug Administration. "The FDA's Drug Review Process: Ensuring Drugs Are Safe and Effective." Last modified November 24, 2017. https://www.fda.gov/drugs/drug-info rmation-consumers/fdas-drug-review-process-ensuring-drugs-are-safe-and-effective.

U.S. Food & Drug Administration. "Priority Review." Last modified January 4, 2018. https://www.fda.gov/patients/fast-track-breakthrough-therapy-accelerated-approval-priority-rev iew/priority-review.

U.S. Food & Drug Administration. "Transfer of Therapeutic Products to the Center for Drug Evaluation and Research (CDER)." Last modified February 2, 2018. https://www.fda.gov/about-fda/center-biologics-evaluation-and-research-cber/transfer-therapeutic-products-center-drug-evaluation-and-research-cder.

U.S. Food & Drug Administration. "What Are 'Biologics' Questions and Answers." February 6, 2018. https://www.fda.gov/about-fda/center-biologics-evaluation-and-research-cber/what-are-biologics-questions-and-answers.

U.S. Food & Drug Administration. "New Drug Development and Review Process." Last modified June 1, 2020. https://www.fda.gov/drugs/cder-small-business-industry-assistance-sbia/new-drug-development-and-review-process.

U.S. Food & Drug Administration. "New Drugs at FDA: CDER's New Molecular Entities and New Therapeutic Biological Products." Last modified January 10, 2020. https://www.fda.gov/drugs/development-approval-process-drugs/new-drugs-fda-cders-new-molecular-entities-and-new-therapeutic-biological-products.

U.S. Government Accountability Office. *Drug Development: FDA's Priority Review Voucher Programs*, January 2020. Washington, DC: GAO. https://www.gao.gov/assets/710/704 207.pdf.

van der Sluis, Justin, Mirjam van Praag, and Wim Vijverberg. "Education and Entrepreneurship Selection and Performance: A Review of the Empirical Literature." *Journal of Economic Surveys* 22, no. 5 (December 2008): 795–841.

Vasella, Daniel with Robert Slater. *Magic Cancer Bullet: How a Tiny Orange Pill Is Rewriting Medical History*. New York: HarperCollins, 2003.

Venter, Craig, and Daniel Cohen. "The Century of Biology." *New Perspectives Quarterly* 14, no. 5 (1997): 26–31.

Verganti, Roberto. "The Myths That Prevent Change." *Harvard Business Review*, May 16, 2012. https://hbr.org/2012/05/the-myths-that-prevent-change.

Verganti, Roberto. *Overcrowded: Designing Meaningful Products in a World Awash with Ideas*. Cambridge, MA: MIT Press, 2017.

"Viagra: How a Little Blue Pill Changed the World." *Drugs.com*, February 24, 2020. https://www.drugs.com/slideshow/viagra-little-blue-pill-1043.

Villiger, Ralph, and Boris Bogdan. "Getting Real about Valuations in Biotech." *Nature Biotechnology* 23, no. 4 (April 2005): 423–28.

Villiger, Ralph, and Nicolaj Hoejer Nielsen. *Discount Rates in Drug Development*. Basel, Switzerland: Avance, 2011. https://pdfs.semanticscholar.org/c91f/88118fd68982b3211ebe6 1b9c6232c1d9983.pdf.

Vogel, Gretchen. "Scientists Take Step toward Therapeutic Cloning." *Science* 303, no. 5660 (February 2004): 937–38.

Wade, Nicholas. "Once Again, Scientists Say Human Genome Is Complete." *New York Times*, April 15, 2003. https://www.nytimes.com/2003/04/15/science/once-again-scientists-say-human-genome-is-complete.html.

Wagner, Joachim. "Testing Lazear's Jack-of-All-Trades View of Entrepreneurship with German Micro Data." *Applied Economics Letters* 10, no. 11 (2003): 687–89.

Wagner, Joachim. "Are Nascent Entrepreneurs 'Jacks-of-All-Trades'? A Test of Lazear's Theory of Entrepreneurship with German Data." *Applied Economics* 38, no. 20 (2006): 2415–19.

Walsh, Gary. "Biopharmaceutical Benchmarks 2014." *Nature Biotechnology* 32, no. 10 (October 2014): 992–1000.

Walsh, John P., Wesley M. Cohen, and Charlene Cho. "Where Excludability Matters: Material versus Intellectual Property in Academic Biomedical Research." *Research Policy* 36, no. 8 (October 2007): 1184–203.

Wanerman, Robert E., and Susan Garfield. "Biotechnology Product Coverage, Coding, and Reimbursement Strategies." In Shimasaki, *Biotechnology Entrepreneurship: Leading, Managing and Commercializing Innovative Technologies*, 2nd ed., edited by Craig Shimasaki, 499–512. London: Academic Press, 2020.

Wei, Dan, and Adam Rose. *Economic Impacts of the California Institute for Regenerative Medicine (CIRM)*. Los Angeles: University of Southern California, 2019. https://www.cirm. ca.gov/sites/default/files/CIRM_Economic%20Impact%20Report_10_3_19.pdf.

Werth, Barry. *The Billion Dollar Molecule: One Company's Quest for the Perfect Drug*. New York: Simon & Schuster, 1994.

Whalen, Jeanne. "Virtual Biotechs: No Lab Space, Few Employees." *Wall Street Journal*, June 4, 2014. https://www.wsj.com/articles/virtual-biotechs-no-lab-space-few-employees-140 1816867.

Whitler, Kimberly A. "New Fortune 100 CEO Study: The Top Graduate Schools Attended by Fortune 100 CEOs." *Forbes*, September 28, 2019. https://www.forbes.com/sites/kimberly whitler/2019/09/28/new-fortune-100-ceo-study-the-top-graduate-schools-attended-by-fortune-100-ceos/#7947f67a2719.

Wickelgren, Ingrid. *The Gene Masters: How a New Breed of Scientific Entrepreneurs Raced for the Biggest Prize in Biology*. New York: Times Books, 2002.

Wilson, Erin, Robert Willoughby, and Mark Wallach. *CRO Industry Primer*. Zürich, Switzerland: Credit Suisse, 2016. https://research-doc.credit-suisse.com/docView?langu age=ENG&format=PDF&source_id=csplusresearchcp&document_id=807217850&seria lid=bWxS47N1G4l%2f5p8umJGkcSVAYjNNuJYDxdjMq6OmgRA%3d.

Wuyts, Stefan, and Shantanu Dutta. "Licensing Exchange—Insights from the Biopharmaceutical Industry." *International Journal of Research in Marketing* 25, no. 4 (December 2008): 273–81.

Wyman, John. "SMR Forum: Technological Myopia—the Need to Think Strategically about Technology." *Sloan Management Review* 26, no. 4 (Summer 1985): 59–64.

Yi, Doogab. *The Recombinant University: Genetic Engineering and the Emergence of Stanford Biotechnology*. Chicago: University of Chicago Press, 2015.

Zhang, Ruiwen, and Hui Wang. "Antisense Technology." In *Cancer Gene Therapy*, edited by David T. Curiel and Joanna T. Douglas, 35–49. Totowa, NJ: Humana Press, 2005.

Zhao, Hangyu, and Zongru Guo. "Medicinal Chemistry Strategies in Follow-on Drug Discovery." *Drug Discovery Today* 14, nos. 9/10 (May 2009): 516–22.

Zhao, Hao, and Scott E. Seibert. "The Big Five Personality Dimensions and Entrepreneurial States: A Meta-analytical Review." *Journal of Applied Psychology* 91, no. 2 (March 2006): 259–71.

Zinner, Darren E., Dragana Bolcic-Jankovic, Brian Clarridge, David Blumenthal, and Eric G. Campbell. "Participation of Academic Scientists in Relationships with Industry." *Health Affairs* 28, no. 6 (November 2009): 1814–25.

Index

For the benefit of digital users, indexed terms that span two pages (e.g., 52–53) may, on occasion, appear on only one of those pages.

Tables and figures are indicated by *t* and *f* following the page number